FASTWALKER

FASTWALKER

(A Novel)

JACQUES VALLEE

In collaboration with
Tracy Tormé

Frog, Ltd.
Berkeley, California

FASTWALKER

First American edition published 1996 by Frog, Ltd.

Frog, Ltd. books are distributed by
North Atlantic Books
P.O. Box 12327
Berkeley, California 94712

Cover art by Sergei Ponomarov
Cover and book design by Paula Morrison
Typeset by Catherine E. Campaigne

Printed in the United States of America

Distributed to the book trade by Publishers Group West

Library of Congress Cataloging-in-Publication Data

Vallee, Jacques.
Fastwalker / Jacques Vallee.
p. cm.
ISBN 1-883319-43-9
1. Unidentified flying objects—Fiction. I. Title.
PS3572.A4129F37 1996
813'.54—dc20 96-57
CIP

1 2 3 4 5 6 7 8 9 / 00 99 98 97 96

NOTE TO THE READER

This book is based on *Alintel,* a French-language novel published by *Mercure de France* in Paris in 1986. In its present form it owes much to my friends Tracy Tormé, who worked on a screenplay version from which I have borrowed large sections of the American dialogue. His creation of original characters and situations that helped the story come to life was an inspiration for our work. I am also grateful to Robert Weiss, who shaped its development, and to Mary Buckley, who edited the text.

Many dedicated people who relentlessly pursue the UFO enigma and related topics have inspired and enriched this tale. Among those who contributed to the fictionalized accounts told in this book are witnesses of unexplained sightings who have communicated with me over the years, and the victims of abductions who conveyed the complexity of the phenomenon and its effect on their lives.

Special thanks are due to my wife Janine, who is cheerfully encouraging my continuing efforts to make sense out of the whole mystery, and to a few friends who, over the years, have helped me reflect on it. Among them I am especially grateful to Fred Beckman for his support and to Richard Grossinger, Keith Harary, John White and "Cloud Rider" on Internet for some stimulating comments.

This is a work of fiction inspired by real events, but primarily designed to entertain. As such, it does not necessarily reflect my own views as a scientist researching the UFO phenomenon. It includes a few specific mentions of well-known real figures like Professor Hynek or Major Donald Keyhoe, and of publicly identified organizations like the CIA and the People's Temple. Apart from such references, any resemblance between characters or groups mentioned in this book and real persons, dead or alive, or with military or civilian organizations is purely coincidental, and did not enter into the authors' intentions.

As to the existence of Alintel, it can neither be confirmed nor denied.

—Jacques Vallee

CONTENTS

PART THREE: EVIDENCE

PROLOGUE: NEW GUINEA

It happened in the Southern Hemisphere, in New Guinea to be precise, on a fine Tuesday evening in the spring of 1998, which of course meant autumn in the United States. The sun had set over the jungle, but the skies were not completely dark yet.

An aborigine with chocolate-colored skin was carefully stepping among the trees in the crimson-purple light. In traditional Papuan style, he was wearing strips of bark wrapped around his hips, as well as a necklace and several colorful bracelets. His nose was pierced by a small stick. As he walked along he paused from time to time to examine the tracks of some wild animal, or to pick up small stones with his toes; he selected a few of them, placing them into a pouch he carried on his belt.

The forest was peaceful; eventually he decided to rest under a large tree, and fell into a deep, restful sleep.

He didn't see the airplane arrive.

In a nearby clearing, the complex structure of a formidable craft, a machine the convoluted military jargon called VTOL for "vertical take-off and landing," touched down with a soft, muffled sound. The menacing device, with its appendages, complicated flaps, variable geometry, bristling antennae and high-tech engines, was alien indeed to the primitive night.

Two weird figures appeared as the canopy opened, wearing dark flight suits, night-vision goggles, communication helmets and bizarre weapons. They entered the jungle with careful, silent steps.

The two men soon came within range of the sleeping aborigine. Through the infra-red scope of a tranquilizing gun he was outlined as a broad target, his bare shoulder clearly exposed. The trigger was squeezed. He collapsed gently on his left side in the tall grass. Hurriedly the Americans carried the unconscious bushman into the aircraft, and strapped him in.

As they took off, one of the men looked at his watch. They had only been on the ground for three minutes and forty seconds.

* * *

"Be careful with this guy," cautioned Joe Wilkinson, the pilot. "You don't want him to wake up and start throwing up all over my equipment."

The plane had reached its cruising altitude, high above the Pacific Ocean; the men were relaxing after the intense excitement of the capture. Greg Parker, Joe's companion and co-pilot, looked more like a post-doctoral researcher in biology than an Intelligence operative. He busied himself inserting a needle and microcomputer-controlled injection lines into a vein in the aborigine's left arm, and connected a vial to the system. Drop after drop fed through the plastic valve.

"Who do you take me for?" asked Greg, nervously tugging at his beard. "I've done this before, you know." But Wilkinson was not joking.

"There's a long history of botched special ops," he said. "We don't need to write our own chapter."

Greg laughed. "Who would ever find out if we did?"

"Of course it's heavily classified," Wilkinson conceded. He went on in mock-professorial tone: "The fact remains, *Doctor* Parker, that a whole generation of psywar experts has often bungled secret deception projects. Our elders were not always discreet . . ."

"Classified or not, everybody's heard about it," replied Greg, who felt a bit piqued every time his fellow officers hinted at information to which he wasn't supposed to be privy. "In my day, every radical on the Berkeley campus was familiar with operation 'Midnight Climax,' when the CIA used prostitutes in a network of bordellos it set up in San Francisco."

That memory brought a smile to Wilkinson's lips.

"Those were the good old days! National Security Whorehouses! Well, boys will be boys, as President Reagan used to say. I wish I'd been there. A voyeur's dream. Agents watched the action through two-way mirrors while the girls slipped LSD into their clients' drinks. Must've been quite a show."

But Parker was not amused.

"Much of that work, all that MK-ULTRA stuff, was plainly irresponsible," he said reproachfully.

"They called it human ecology. They were looking for a good interrogation drug, a workable key to memory and the doors of consciousness," Wilkinson went on. "Who could blame them?"

"Consciousness my ass. There wasn't a bit of useful science in it. Likewise, when Army Intelligence sprayed a solution of anthrax in the New York subway. Do you think everybody has forgotten that?"

"Yes, as a matter of fact, I do believe people have forgotten it," the pilot answered coldly. "Or they don't care, which amounts to the same thing. I also believe you haven't studied the material well enough to know what you're talking about."

"Well, what about the time the Air Force flew specially-equipped planes to disperse microbes responsible for various kinds of respiratory ailments off the coast of San Francisco? They allegedly needed to study the possible effects of Soviet biological warfare. How many elderly folks died from acute pneumonia in the following weeks? Do you think the families have forgotten that?"

"Come on, Greg, you're a scientist! How could you prove that the effect was statistically related to the cause?" Joe asked with a chuckle. "Anyway, the lessons from these operations have been learned well," he added in a reassuring voice. "We don't take that kind of chance any more."

Greg knew that, in this instance at least, his partner was telling the truth. He had received his Ph.D. in anthropology from the University of California at a time of intense student turmoil, and he certainly understood the importance of utter secrecy in cross-cultural operations. Besides, the outfit they both worked for, known only as Alintel, was much more secret than either the CIA, the Air Force's notorious Office of Special Investigations, or Army Intelligence, which they regarded at best as narrowly specialized operatives or, at worst, as a hopeless bunch of amateurs. Alintel was so secret, in fact, that the codeword for the project had been made an integral part of the project itself, so that it was "recursively unfindable," as their leader, General Bushnell, used to say with a mischievous smile. There was no way to even know that the team existed if you hadn't specifically been selected to be part of it.

* * *

The plane reached an Air Force base in Southern California on Wednesday afternoon. The sun was already scorching the tops of the palm trees, and the runways, which had been cleared for this high-priority incoming flight, appeared hazy, shimmering in the heat. The aircraft taxied to the large metal door of a hangar. A team of specialists were awaiting it. The unconscious aborigine was carried from the plane, strapped into a wheelchair, and rushed inside. The door closed behind them with a clang.

They found themselves in a vast studio with a wide screen, a Dolby stereo system, and a console with many cables. Greg replaced the vial that fed into the man's vein.

The aborigine opened his eyes in utter surprise. The flow of drugs into his bloodstream was carefully programmed. Recordings from sensors attached to his frontal and temporal areas, his chest and his genitals were beginning to register his reactions as he heard the opening theme of *Star Wars* blasting from all sides at full volume.

The bushman was intrigued, fascinated, petrified, excited and terrified as the movie toyed with his emotions. He perspired, trembled, tried to close his eyes, but the drugs heightened his attention to the breaking point. He watched the hologram of Princess Leia dancing on the screen. He saw Darth Vader pulling out his laser sword.

The images brought him beyond terror. The man urinated on the spot and rapidly fell into unconsciousness. Behind him, a bank of recorders whirred softly.

"I'm switching medication again," Greg spoke into a microphone, for the record.

The straps were removed; they wheeled away the chair.

The whole team came to attention when an older officer strolled onto the set. A tall man with graying temples, blue eyes and thin silver-rimmed glasses, Cliff Bushnell wore the uniform of a four-star Air Force general.

"I'm going to need full data acquisition on the cultural re-insertion of this individual," he sternly reminded the team as technicians busied themselves around their subject.

Soon the native's ornate necklace featured a new device, a tiny transmitter whose signal could be conveniently picked up by the Milstar satellite constellation.

It was just past sunset when Greg and Joe landed again in New Guinea with their passenger, at the precise spot where they had touched down the day before. All the medical instrumentation was unhooked from the unconscious native. Only a tiny puncture mark was left: They masked it with a drop of collagen. The infrared scanners showed no sign of human presence in the vicinity of the site.

Carefully, almost respectfully, the two agents carried the sleeping bushman to the foot of the very same tree where he had been picked up. They took off again, and turned the nose of the powerful craft towards California.

* * *

"Anyone who didn't believe that we could accomplish this mission in less than twenty-four hours using today's off-the-shelf technology hasn't paid attention to scientific developments in the last twenty years," said General Bushnell as the operational debriefing got under way. "And anyone who doubts that such experiments have in fact been conducted since World War II without the public or Congress being aware of them is woefully out of touch with reality."

"What's reality, anyway?" asked Dr. Beverly Bernard with a chuckle. A blonde, short-haired woman in her late thirties, she wore a blue suit adorned with a rectangular plate bearing her name, but her face was a mask. An expert in the chemistry of neurotransmitters, she held a doctorate in medicine and a Ph.D. in biology from Northwestern University.

Greg butted in. "There's a simple answer to that, in this particular case at least." He looked around at his colleagues. "Let's imagine that we could watch our aborigine as he wakes up in the jungle, in his everyday, normal world of Papua, the world we could call *Reality One*. This man has had an experience—a vivid, life-changing experience, a profound catharsis.

He is aware that he has lived, however briefly, in another reality. Let's call it *Reality Two*. He may not have understood the words, but the images were vivid enough. He will remember most of it: the Death Star, Princess Leia, Darth Vader's laser sword. For the rest of his fucking life."

"So he rushes to his village to tell his breathless tale before the assembled elders ..." commented Joe Wilkinson. "But he has no access to our own reality, the technology that made it all happen."

"*Reality Zero*," joked Greg Parker. "The basis for everything. The plain reality of engineering. Sprockets; celluloid; Dolby decoders. No, he has no access to it. As far as he knows, it doesn't exist at all. *It is unthinkable*."

"Unthinkable or not, he'll have to explain a few things to his friendly local medicine man," broke in Beverly with a guffaw.

"So what?" Parker went on. "The medicine man won't need to examine him very long to convince himself that the fellow shows all the signs of terror, genuine terror: His pulse, the look on his face, the sweat on his forehead, the smell of his skin, his broken breathing will all speak most eloquently of his obvious fear: He recalls what he saw and heard. The whole episode, this alien trip, this experience, is indelibly etched into this man's memory forever."

"And that is the reality we control," concluded General Bushnell with a satisfied look around his hand-picked, expert staff. "The world has changed, my friends. In this new political landscape of ours, it's the only reality worth controlling."

PART ONE

ALINTEL

Fastwalker: in the jargon of scientific intelligence, a source of electromagnetic radiation moving at high speed in the outer layers of the atmosphere, which triggers the sensors of spy satellites.

CHAPTER ONE

Rachel Rand was running late, and it was well past five-thirty that fateful Saturday when she finally found the right street in the fashionable but convoluted Long Island neighborhood where the birthday party was held.

Everything about Evelyn Manchester's house reminded Rachel of a sailing ship in distress. The oddly slanted roof that tumbled towards the garage contributed to the sinking appearance; the great white columns of the portico looked like masts, and the drapes that flapped in the wind through the open windows seemed as hopeless as torn sails. The general disorder that reigned all around, the screaming kids, the front door draped with a banner that read: *"Happy Birthday, Norman!"*; the toys strewn about on the porch and the screams of joy or anger that emanated from the structure did nothing to dispel this impression of impending chaos.

Rachel climbed up the front steps, urging her niece and nephew ahead of her. She rang the bell, but no one heard or cared, so she pushed open the door. The Manchesters' living room was rocking to the music of Madonna. Two dozen kids dressed as little Batmen, Jokers, and cartoon characters had turned the place into complete bedlam.

Norman's mother finally made her way through a crowd of gyrating fourth-graders, and spotted the new arrivals. Rachel Rand and the two kids offered an interesting contrast. Rachel was in her late twenties. She had wispy light-brown hair and bright blue eyes. She was pretty, with an intelligent, compassionate, expressive face, and wore a sensible business suit; but the two children were another story: Her eight-year old niece was dressed in torn leotards and a leather jacket, while the six-year old boy wore a brown suit that was two sizes too big, a gaudy red tie and a thick, off-center hairpiece.

"Mark, Vicki, I'm so glad you could come!" the weary mother said. She turned to the young woman with an inquisitive smile: "And you are . . . ?"

"I'm their Aunt Rachel, Rachel Rand," she replied.

"Oh, you must be Bonnie's sister, the paralegal, right?"

"I'm with Robertson, Benraph and Chernik," she asserted, justly proud. They were among the rising stars of the financial district.

"Won't you come in and join us?"

Before Rachel could answer, ten-year old Norman burst in. The birthday boy was wearing a Ninja Turtle costume, and had fully entered into his character. He began to circle his mother aggressively, kicking her in the shins, martial-arts style, with his mutated turtle feet. Rachel covered her mouth, trying not to laugh as poor Mrs. Manchester pretended to ignore her marauding son by talking to Mark.

"My, what interesting costumes," she said, wincing with pain. "And what are you supposed to be?"

"Ted Koppel," answered Mark reluctantly. "It was Dad's idea," he added with a frown.

"And I'm Paula Abdul!" stated Vicki with all the assurance of her age.

Before his mother could respond, Norman swiped Mark's hairpiece, making him scream. A glass shattered somewhere inside. A boy dressed as a cowboy managed to lasso a tall vase on a side table, sending it crashing to the floor. Pandemonium broke loose throughout the house.

"I'll call to check up on them in a little while," Rachel stated as if nothing had happened. "Their Mom and Dad'll pick them up later this evening."

She realized that if she were to get away safely, now was the perfect time, so she retreated from the shipwrecked mansion as gracefully but as fast as she could manage. As she got into her green Volvo, she caught a last glimpse of Norman's frazzled mother weakly waving at her from the front steps of the house.

"Sure you don't want to stay awhile? There's still plenty of cake," the lady offered.

"I'd love to," Rachel lied as she started the car, "but I've got some urgent paperwork waiting for me back home."

"Thanks again, and good luck!" She shouted into the wind, in an effort to be heard over the cacophony of screaming kids as she pulled away—just before Norman found the firecrackers.

* * *

Dusk was falling and Rachel was driving down the two-lane road that ran into the woods. She was listening to the radio and musing about destiny. Kids were wonderful, she thought, picking up loose candy wrappers from the passenger seat to stuff them into the ashtray, but she wasn't quite ready for a family of her own.

There was no man in her life at the moment, not even an occasional reunion with former boyfriends, and she had fought successfully the attentions of some of the young attorneys at the law firm. The lack of romance had its quiet, pleasant side. A couple of years earlier, after an unlucky love affair with a married man, she had gone through a rough emotional time. She had even rushed out and applied to the Peace Corps, hoping they would send her to some faraway land where she could help others and heal herself. The woman who interviewed her had wisely recognized her motivation as a passing storm of the heart, and turned her down.

No longer mortified at the rejection, Rachel was happy with herself. She took time to enjoy every day, as she was now enjoying the drive in the fine November evening, listening to a Kitaro tune, admiring the golden Fall foliage on both sides of the twisting road. She would get home in time to light a fire and curl up in front of it with the book she had started to read, a biography of Isabelle Eberhardt.

Everything about Isabelle Eberhardt was as extraordinary as Rachel's life was simple. A beautiful, feminine figure, she always dressed as a man. A European, she had turned Arab. A woman of Russian origin, she was torn by what the author called "untidy mystical torments." Born in the prim and proper city of Geneva, she had died in the middle of an African desert—drowned by a sudden flood! She was an outcast, but she left no one indifferent. Her love affairs were scandalous ... Rachel couldn't wait to get home and read about that legendary woman. She caught herself looking at her reflection in the mirror, wondering if those blue eyes of hers could have held the attention of Arab cavaliers; she imagined the warm wind of the Sahara in her hair.

She thought nothing of it at first when the music was marred by static. This wasn't unusual in the countryside. Stranger was the fact that she found the same static everywhere when she switched stations. Puzzled, she fiddled with the dial, but the sound was unchanging, as unpleasantly urgent as oil sizzling in a frying pan, even across the AM band. Rachel looked around for a microwave tower, a transmission line, even railroad tracks that might have caused the interference, and saw nothing.

In her headlights the woods were beautiful, quiet and lonely.

Too lonely.

Rachel was breathing hard, but she became very still when a brilliant flare, like a sustained lightning flash, suddenly illuminated the inside of the sedan. Something big passed, close to her, just over the top of the Volvo.

She looked up and saw a huge metal disk. The engine died.

Rachel found herself gasping for breath as the car rolled silently down the road. She hit the brakes, bringing the vehicle to a jarring stop, and she frantically tried to restart it. The engine gave off a series of stifled coughs, but didn't come to life. Still turning the key, she looked up the dark highway again and saw an incredible sight: The glowing disk was silently descending over the road and kicking up dust, about twenty-five yards ahead. It was hovering a few inches above the asphalt, gently rocking to and fro. It spanned both lanes.

In mounting panic, Rachel struggled again with the ignition. The engine briefly sputtered, but still refused to turn over. Rachel kept trying, praying to God in a trembling whisper. She dared to glance up the road again. A wall of light was drifting between the object and the car. It seemed that someone was drawing a white curtain of undulating fog in front of Rachel. And now several goblin-like figures with oddly oversized heads were walking toward the little car.

Rachel felt frozen, like a deer caught in approaching headlights. In a motionless frenzy, she tried to reach around the car. She wanted to lock the doors, but was shocked to discover that her arm—her whole body in fact—had become warm and heavy, too heavy for her to complete the motion. Physically, it wasn't an unpleasant sensation, rather like sinking blissfully into a tub of warm water. Psychologically, it was terrifying beyond the limits of horror.

The little men surrounded the car. They pressed their misshapen chalk-white faces against the glass. Hairless heads dominated by huge, depthless, sad black eyes seemed to be everywhere. One of the grotesque dwarves appeared to be the leader. He pointed a glowing rod-like device at her. Rachel screamed.

* * *

Another hour and he would be home, Higgins thought, as he geared down to move the truck around a twist in the darkened road. It had started raining and he had trouble seeing the poorly-marked center line through the rhythmic sweep of the big, noisy wipers that kept a steady beat to the song on the radio. A burly man, Higgins controlled the eighteen-wheeler with a sure hand. He was hauling a load of plasterboard towards the city, along a road he knew well.

The truck entered a straight section of the highway. Higgins reached for a doughnut, but before it made it to his mouth he saw the car parked in the grass. He glimpsed the figure of a girl in the middle of the highway. He slammed on the brakes. "Jesus!" he cried. The truck began a slow, agonizing skid on the wet blacktop, tires screaming, horn blaring.

The woman didn't budge.

Through his windshield Higgins saw her standing there, unflinching as the big-rig bore down on her. The moaning truck barreled her way, fishtailing as he desperately tried to stop before it was too late. It finally skidded to a halt, inches from the unblinking woman.

Higgins jumped out, ready to yell at her furiously. But he felt his fury vanish when he saw her blank expression, the unearthly emptiness of her gaze. He moved toward her slowly, spooked by her strange stillness, while rain pelted down on both of them, making iridescent puddles and bouncing on the pavement.

Rachel was soaked to the bone, yet she didn't even seem to notice the storm.

"My car won't start," she said simply, softly.

"Lady ... that's a hell of a way to flag down a truck," Higgins replied, the adrenaline draining from his blood. "Let me see what's wrong."

He parked his truck behind Rachel's Volvo, on the shoulder of the road. The first thing he saw when he peered into the car were the keys dangling from the ignition. He reached over and gave one turn. The engine came to life at once. He jumped out, puzzled.

"You told me it wouldn't start," he said, unnerved, with a tone of bitter reproach.

"It ... it wouldn't," she replied blankly.

She seemed disoriented, confused, disturbed. Higgins took one step back and looked at her.

"I'll be al-alright," she stammered. "Thanks for your help."

He watched her get behind the wheel of the Volvo. Blood slowly began to trickle from her nose. Spooked, Higgins stepped away, climbed into his truck and got the hell out of there.

* * *

Rachel lived in a modern highrise near LaGuardia airport. It was a relatively safe area, and she felt a little better when she reached the fourteen-storey building with its up-to-date security system, its brightly-lit lobby and its bank of well-kept elevators. Yet the first thing she did when she entered her small apartment was to turn on all the lights. She locked the door carefully, and drew all the drapes.

She was still shaky and dazed when she took off her drenched coat with slow, mechanical gestures. She stepped into the living room. When her cat meowed a greeting, she jumped nervously at the sound.

Rachel took a deep breath and tried to gather her thoughts. Then she suddenly remembered the kids! She was supposed to make sure they were alright! She rushed to the phone. She dialed a number, and her elder sister Bonnie answered after a few rings, her voice a muffled groan.

"Hello?"

"Bonnie? It's Rachel. You—you sound like you were sleeping?"

"Of course we're sleeping, dumb-dumb," said her sister's voice. "What's the problem?"

"Bonnie, what are you doing?" insisted Rachel, alarmed. "You're supposed to pick up the kids from the birthday party."

"The kids are in bed—hours ago. What's wrong with you?"

"But I—I just dropped them off . . ."

"Rachel, it's two A.M. What's wrong with you?" Her sister asked again.

Rachel went pale at the sound of "two A.M." She grabbed a desk clock and stared at it in stunned disbelief.

"Oh, God, I need a cup of coffee," Rachel said weakly.

"Good idea. Look," Bonnie went on with some irritation in her voice, "I don't know what your trip is tonight, but can't we talk about it in the morning?"

Frozen with fear, Rachel simply let the receiver drop from her hands. She began to shiver, rubbing her arms as though they were suddenly soiled.

* * *

For a long time Rachel stood in the shower, feverishly scrubbing herself like a rape victim trying to cleanse her body. She sobbed, her tears lost in a torrent of hot water.

After a long time she put on a white bathrobe, dried her hair, sat down in the living room and tried to calm herself by breathing slowly, exhaling with deliberate intensity. Finally she got up and made herself a cup of coffee; then she opened the phone book and dialed the police.

"So, where's your car now, lady?" the desk sergeant asked after she had explained her story as succinctly as she could.

She heard the boredom in his voice. "They must get calls like this all the time," she thought. "Lonely women who wake up in the middle of a nightmare, and don't know where to turn."

"It's in the garage, of course," she said, trying to sound professional, very matter-of-fact.

"Is there any evidence on the vehicle itself, Ma'am?" He must have been filling out a form, because she could hear some paper shuffling in the background.

"I . . . I don't think so. You mean dents, or burns?"

"Yeah, peeling paint, stuff like that?" He tried to sound helpful. It occurred to her that he may just have been keeping her talking, in order to assess whether she was disturbed enough to do something foolish—try to kill herself, or create some sort of disturbance.

"No peeling paint, nothing like that," she replied.

"And what about you, are you alright?"

Before she could answer he added: "Have you had anything to drink since the alleged incident, Ma'am?"

"Just coffee."

He sighed, the tone of boredom returning.

"We'll advise you if we get any other calls, Ma'am. In the meantime feel free to come in to the station during office hours and fill out a full report. For the record, you know."

She thanked him and hung up, eagerly returning to the phone book. She dialed another number.

"University Center," a young male voice answered. "Observatory."

"I want to report a UFO," Rachel said firmly.

"Oh . . ." He was clearly taken aback. "That's not my department," the man fumbled. "Have you tried the police?"

"Naturally. They've requested that I file a complete report with them," she said, trying to sound official, scientific, in control. "I just thought you might have seen something too."

"We're not observing tonight," he answered with a hint of apology mixed with irony. "You might have noticed it's raining pretty heavily. I'm just here taking some calibration measurements and developing some plates. We're pretty backed up . . ."

The University didn't care if she had been attacked by ugly aliens who looked like oversized gray insects. She would simply be an embarrassment to them.

"But can you explain to me what it is I saw?"

"Well, as I told you, it's not exactly my department. Come to think of it," he added, the irony returning, "we don't even have a department of ufology, you see, so . . . I'm sorry I can't help you."

She pretended to thank him, and put the receiver down, feeling lonelier than ever. She recognized the signs of anger and despair rising within her. Hoping to snap out of it, Rachel turned on the television set. The voice of what sounded like a fire-and-brimstone preacher came booming out, but he wasn't not talking about God.

"I'm speaking directly to those of you who have been TOUCHED," the man said with assurance and urgency. "I don't give a damn about the others—the sleepers, or the skeptical eggheads. I am not speaking to the masses of the incredulous, of the unaware. I speak only to those who have been CHOSEN, those who know our cosmic brothers, who feel the weight of the coming apocalypse." His powerful voice conveyed certainty, anger, urgency.

It forced Rachel to pause, still in a disturbed funk yet intrigued by what she'd just heard. A man's round, bearded face literally filled the screen in an extremely tight close-up. The subtext in pulsating gold letters read: "Dr. Helmut Borodine, Ph.D., Abduction Expert."

The camera backed away slowly, revealing a lectern, then a large room. But what really captured her attention was a shot of faces in the background. Eager, honest, clean faces. They belonged to well-scrubbed young people dressed in blue and purple robes, seated behind rows of telephones. They listened with deep reverence to Borodine, who stood on a high platform. A telephone number started blinking in a corner of the screen.

She went into the kitchen to pour herself another cup of coffee, and she carried it back in front of the set. The angle of the camera had changed.

"*I know who you are,* every single one of you," Dr. Borodine's voice boomed on. "I know how you feel. *Isolated!* Cut off from the rest of society, misunderstood even by your closest family, *because you are special.*"

In spite of herself, Rachel felt the weight of his words; she turned up the volume.

"Come to me," said Dr. Borodine, as row upon row of his disciples nodded assent in the background. "Come to the shelter of knowledge, come to *The Brotherhood of Galactic Science,* a support group for those who have been touched by the Light. I can help you understand your cosmic contacts. You are unique—you are indeed remarkable—but in these End Times, in the Times of the Eschaton, *you are not alone.*"

Rachel was moved. A solitary tear, a tear of hope, rolled down her cheek. In the distance, she barely heard the muted rumbling shudder of the storm. With a trembling hand, she wrote down the telephone number of the Brotherhood.

CHAPTER TWO

The late afternoon was dragging down the autumn sun, casting an orange glow around the desert mountains of Nevada. It was the same Saturday, about the time that Rachel Rand was almost run over by Higgins' truck after the birthday party on Long Island. The jagged, sharply-delineated hills looked forbidding to the two pilots who had been ordered to criss-cross the wispy clouds high above the Earth, in search of an elusive target that was troubling the ground-based radar at Nellis Air Force Base.

When Captain James Hall caught the reflected glint of the glass towers of Las Vegas, very far away, they reminded him of home, of his earthbound family. His mind briefly drifted to his next leave, when he would be back among normal folks. Beyond the sandstone hills the horizon radiated with brilliant streams of purple and golden light. But the sky was empty except for the two steel-blue F-16 jet fighters blasting away like hungry sharks.

The two pilots were so close to each other that Hall could see the angular face of his wingman, Lieutenant Steve Grodins, half-hidden by his slate-black visor.

"Zipgun Zero-One to Nellis Control," he called into his mike, his urban Texas twang colored by a sense of frustration.

"Nellis here. Go ahead, Zero-One."

He recognized the voice of Anne Lynch. He imagined her seated behind the control tower screen where green shapes were swirling on a dark window, and he thought again how much he would hate to have a job like hers, attentively watching a funny-looking display all day long. He had wanted to fly since he was three years old.

"Negative on visual or radar contact," he continued. "Request current designation and location of target."

The answer came back immediately: "Target bearing one-zero, eight miles north-northeast of the range. Altitude one-one-five thousand and descending. Current designation . . . unknown."

So the radar was still tickled by something physical, and that something was literally falling out of the sky. Falling into their waiting hands.

"Understood," replied Captain Hall. "Vectoring to target area."

* * *

Sergeant Anne Lynch, a slender woman in her thirties, was carefully following Captain Hall's maneuver as she monitored her screen, ignoring the quiet chatter of the men around her. They wondered what could be descending from such a high altitude, and so steadily, too. It was not a meteor, nor a re-entering satellite: Both would already have disintegrated, leaving fiery trails across the sky.

A phone by the side of the radar screen rang loudly, startling everyone. Major Garcia picked it up.

"Yes sir, we're monitoring it now," he said. "That's affirmative: Object is anomalous, unresponsive. We've diverted two Sixteens . . ."

A pause, then the Major cut off, his face concerned. Taking a pen from his jacket pocket, he scribbled down a series of numbers on a piece of paper.

"That is highly unusual, sir. Understood . . . Yes, sir . . . At once, sir."

Garcia hung up and spoke urgently into the live mike that connected the tower to the two aircraft.

"This is Nellis Control," he announced. "We have just received new orders. We are now in state Mike X-Ray. This is not a drill, understand? *This is not a drill.*"

Around him in the control room the men were now standing at attention, and no one felt like chatting any more.

* * *

Seated in his cockpit, Captain Hall reacted to the new instructions with a start. He reached for a booklet filled with vertical rows of letters and numbers, and propped it up against a kneeboard.

"Roger . . ." he said after some hesitation. "Uh, please authenticate condition Mike X-Ray."

There was no hesitation in Garcia's reply: "Authentication code: Bravo-Foxtrot-Whiskey-November. Your orders are to switch to Charlie-Yellow-Tango."

"Authentication confirmed," said Captain Hall, still tense. "Switching to Charlie-Yellow-Tango."

For the next few seconds both pilots were busy switching to the classified frequency on their communication gear. The voice of Major Garcia was drowned in high-pitch static, which died down as a new, powerful voice came over the system. Neither one of the pilots recognized the speaker.

"Zipgun Zero-One, *we have taken command of this situation.* Zero-Two will break off search and return to base. Captain Hall will close on unknown, film and identify."

Those were puzzling orders, thought Captain Hall. Why were they sending the other plane back to base? And why did the voice call him by name? Hall looked over at his wingman, who was vehemently shaking his head, and pointing upward.

"I'll see you on the ground, Steve," Hall said. "You heard the instructions."

Reluctantly, the wingman nodded, and the other F-16 rolled into a graceful right-hand bank, breaking down and away. Hall kept climbing, alone.

Suddenly, "I have radar contact!" he yelled. "Target reads stationary at seven-zero thousand. I'm climbing to establish visual."

"Captain Hall, your ceiling is six-zero thousand."

"I'm aware of that, sir. But I can't film what I can't see."

There was no answer to that statement, so Hall continued, craning his neck, scanning the sky, the plane shuddering around him in the severe ascent into the thin air. Hall's head felt heavy and his breathing became difficult.

A glow, then a faint flash, reflected off his visor.

"Six-zero thousand," he said slowly, "I see something."

"Zero-One, can you describe the target?" asked the ground controller.

"Need ... to ... get closer," the pilot managed to say.

And he thought, "What is wrong with us, that we automatically call any object in the sky a *target,* as if we had to shoot down anything we don't understand?

The plane was vibrating wildly, and he fought to maintain control, straining to see the object more clearly. It was a strange silvery thing, hanging there at the top of the world. And it seemed to be undulating. Could it be something as simple as a scientific balloon lost in the high atmosphere, rippling in the jet stream?

There was a loud cracking sound. Sparks spurted out of the front and the back of the jet fighter.

The engine went dead.

Now there was no more rumble in the cockpit, no vibration, just the impression of the wind rushing around the canopy. Captain Hall frantically tried to restart the engine. The aircraft gracefully reached the top of its climb and started the descent, plummeting back towards the ground.

"Zipgun Zero-One," he gasped urgently, "I have compression failure. Full loss of power. No warning bells, nothing."

The controllers answered with words Captain Hall did not understand. The sky and the ground were spinning around him in an ever-increasing spiral. He came close to blacking out, struggling with the stick as much to force himself to remain conscious as to regain control of the diving jet. But the aircraft continued to fall like a stone. Hall breathed hard, frantically pressing the ignition button, getting no reaction.

"Come on, baby ... come on ..."

Too late to bail out: The desert was looming large ahead of him. Hall glanced at the wings, wondering when they would start shearing off under the tremendous stress, and pressed the button one more time. An alarm rang, there was a faint sputter, and the big engine kicked in, jolting the pilot in the back.

"Holy shit!" yelled Hall, pulling the plane out of its dive.

But before he could relax he found himself inside an ominous dark shadow. Something very large loomed above him and rushed by. It was shiny and metallic, and shaped like a disk.

"Zipgun Zero-One," he said breathlessly, "I have restart. Repeat:

I have restart. And I have company."

The object made another close pass, streaking by with a whoosh. Hall threw the F-16 into a roll, and again the silvery disk mirrored him, passing very close. He activated the gun camera, but couldn't get the object in his sights.

"It's dropping now!" he informed the ground controllers. "Skimming over the sand dunes ... Will try and film it."

He leveled off a few feet from the desert floor, flying past occasional rocks and shacks as he raced after his target.

"Come on, you son of a bitch, smile for the camera!" he urged, locking it in the double circles of his collimator, "that's it!"

But his exuberance was soon tempered by the sight of a jagged mountain range, with a high peak shaped like a huge pyramid, coming up in front of him.

"Jesus! It's gonna crack up! It's gonna burn!"

He continued to film the thing in front of him, and braced for the explosion as the disk reached the mountain at top speed. But there was no explosion, just a glow, an intense display of light, turning to a brilliant series of colored beams that burst into the sky. They came from a point at the foot of the mountain range, and extended into the atmosphere with a shimmering effect.

The beams encircled the object.

Captain Hall had no time to reflect on what he had just seen, because he was coming up fast on the rocky range himself. He pulled hard on the stick and barely skimmed over the peak. Exhaling with relief, he looked up and saw something that astounded him: There was no sign of the target he had been pursuing. It was as if the craft had been swallowed up by the mountain. For a second he even thought he had seen the rocky slope open up into a gigantic cavern, and close up again.

"No way, no fucking way," he whispered.

Hall kept his aircraft level, and continued his flight in stunned silence.

* * *

The ground was dark when the F-16 descended. Hall did not recognize any features. He scanned the desert with growing concern.

"Zipgun Zero-One . . . Am approaching the coordinates you gave me," he said. "But there's nothing on my chart . . . nothing down here but desert."

The answer was immediate, terse and formal.

"Prepare for landing," came the voice of the ground controller. "Lower the gear. Do what you're told."

Panels swung open underneath the jet, and the powerful wheel assemblies dropped, their shock absorbers extending like the clawed legs of an eagle. The jet was only a hundred feet above the sand dunes, and there was still no runway in sight.

"Hell, you want me to land on a goddamn . . ." he began to protest.

He did not finish his angry remarks: blue landing lights had suddenly illuminated the barren landscape, flanking a huge runway camouflaged in the dunes.

A few minutes later the aircraft was on the ground, rolling to a stop at the foot of a rocky hill. Cold-eyed uniformed men rushed towards the jet, unhooked the tail chute, surrounded the canopy and removed the gun-camera cartridges. In the crude light of the hidden air base Captain Hall saw that they wore unfamiliar blue uniforms. They took him down a featureless corridor that was cut into the hillside.

He smiled to the athletic men in blue berets who guided him.

"You guys are Army? Delta Force? Never heard of Blue Berets before," he said in a conciliatory tone. "Who you fellas working for?"

He didn't get a reply. They never even looked at him.

* * *

Hall was sitting at attention in a high-backed chair. In the twenty-four hours since he had landed his F-16 in this remote corner of Nevada something very drastic had happened to him. Sweat drenched his forehead. He was tired and haggard, his eyes bleary, and his face showed a scratchy growth of beard. With the bright light shining on him, he could not distinguish the rest of the room very well. He could barely discern the men who were interrogating him, asking him the same damn questions for the hundredth time.

"I'll say it again, sir, *it went right through the mountain.* I saw it. *Saw it . . . change shape too.*"

"That means it wasn't a solid object, surely you realize that?"

"Yes, sir," said Hall in a tired voice, almost a whisper.

"Is that your position, then, that this phenomenon you saw was not materially physical?"

"Yes. Maybe. I don't know."

"Yet you stated it looked *metallic,* did you not?" This from another interrogator, who spoke English with a strong accent. Hall could not see the man who said those words.

"It did, sir, at first." His own voice was almost inaudible.

"You must be thirsty, Captain," said a friendly man whom the others had addressed earlier as "Doctor." "Why don't you drink this? It will do you some good."

Hall eagerly took the glass filled with what seemed like orange juice, or perhaps grapefruit juice. It was yellow, somewhat foamy. It smelled like pine trees. It had a pleasant taste.

* * *

Reality dissolved around Captain James Hall. He drifted in and out of consciousness, vaguely aware of people looking at him, peering into his pupils, screaming at him, then hugging him like a long-lost friend, for no apparent reason. He slept for a while.

He woke up again, staring at a wall of fluorescent white light. Silhouettes occasionally blocked out some of the glare. One of his new tormentors was a young bearded man in a turtleneck sweater, another was in a Navy officer's uniform, and there was a blonde woman with short hair, who wore a blue suit. He sensed others as well, standing just beyond the light source.

"We have reviewed the flight data," said a female voice, the words impeccably detached and precise. "You will be relieved to learn that the radar echoes were false returns, caused by a temperature inversion," she went on. "That is extremely common in the desert. You obviously hallucinated because of oxygen deprivation. There was nothing up there."

Captain Hall said the first thing that came to his mind.

"The film. It must be on the film."

"The film has been processed and analyzed," answered the cold voice of the man Hall had come to regard simply as "the Beard."

"It shows nothing unusual, even after careful computer enhancement, frame-by-frame."

Hall had to fight a peculiar dizziness that was spreading through his brain.

"I don't believe that," he managed to utter.

The Beard's voice replied: "You disregarded standard procedures, Captain. You recklessly flew over your ceiling, causing compression failure. You nearly destroyed an extremely expensive aircraft in the process. *Those are the facts.*"

His mind reeling, Hall sighed and reached up to touch his burning forehead, to shield his eyes from the lights . . .

"Hands down, Captain!" interrupted the female interrogator. "You are sitting at attention here!"

"What do you want from me? You want me to say I never saw it? Is that it?"

Wearily, he lowered his head. There was silence in the windowless room, and he became aware of the whir of the air conditioners.

"We want you to do the right thing," someone finally told him, in a tone that was flat, almost mocking.

* * *

The bright lights went off. Captain Hall drifted into a state that was on the borders of sleep. He rubbed his eyes. He was in another featureless room, much smaller than he had expected. The bearded man stood next to him, a piece of paper in his hand.

"Sign this, Captain, and you're free to go."

Hall tried to read it, but his eyes were so watery that all he saw were dancing colors, jumbled lines.

"I'll save you the time," said the man. "It states that you will never discuss any element of this non-incident with anyone. Including, naturally, any knowledge of this facility. National security. I'm sure you understand. Do yourself a favor and sign this."

Hall felt anger come to the surface.

"I just wanna get the hell out of this friggin' place," he said as he grabbed the piece of paper and tore it in half.

"Of course you do."

There was no anger in the voice, no emotion.

Carefully, without another word, the man picked up the torn document and headed for the door.

"Hey!" Captain Hall called out.

The man stopped in the doorway. Hall tried to rise, his tired voice betraying a stubborn defiance.

"That was a real flying saucer, wasn't it? *Wasn't it ?*"

The other looked at him compassionately but did not reply. Hall let out a long sigh.

"Damn," the pilot went on. "I didn't think people saw those things any more."

"They don't," said the bearded man as he left, shutting the door behind him.

* * *

"If the Fastwalker had been detected first in Mongolia or over the North Pole, things might have gone very differently," said General Bushnell, gently tapping the huge map that occupied an entire wall of the vault. "Perhaps the Russians or the Chinese would have grabbed it first."

He went on while the others listened intently.

"Historical opportunities are not engineered. They happen. Either you seize them or you don't. It will remain a matter of record that the Fastwalker chased by Captain Hall triggered an alert over Dreamland, and was found to be on a trajectory conducive to capture. This is what Alintel was designed to do. We've been waiting for this opportunity since 1947, with very limited success. This is not another bit of metal, my friends, not the kind of fragment we've retrieved before. This is our first capture of a complete craft."

He fiddled with the pencil and the pad in front of him, and let the words hang over the group.

"The problem before us now is to hide it well enough, and long enough, to exploit this bit of good luck to the fullest extent. I want us to watch the gun camera film again."

Joe Wilkinson rewound the projector, and a few seconds later the images started moving again over the screen.

"It's a type-two Fastwalker," the woman's voice stated in the darkened room. "Like the one the French tried unsuccessfully to intercept in Saint-Geniez in the seventies. No doubt about that."

"Yeah, Beverly is right, watch how it plays with the jet . . ." began Joe Wilkinson.

She interrupted him: "And yet we were able to capture it this time. Our first complete capture in fifty-one years."

The general took a deep breath and motioned for his assistants to stop the film. The lights came back on.

"Any ideas about a suitable cover?" Bushnell asked again, sitting down at the head of the massive conference table.

"Sir, we could spread another crash story," suggested Greg Parker.

Wilkinson added helpfully: "Yeah, sprinkle some exotic material around the desert."

"A ceramic matrix . . ."

"That would lend credibility to the theories we've been promoting," added Dr. Beverly Bernard approvingly. But the general deflated her enthusiasm.

"That'd take too long. We need to start assembling the research teams. Denial is not an option this time. You only have two choices: openness, or disinformation."

"Sir, I may have a solution," announced Greg Parker. "Some years ago our esteemed Soviet colleagues fooled the CIA by planting UFO stories among their own public. They were launching satellites that violated the SALT treaty, as you may recall, and the whole population could see the boosters going up. But our experts only heard the UFO tales. Back in Langley the analysts put it down to ridiculous nonsense, so they didn't bother to check further."

The general shrugged impatiently. "What's your point ?" he asked.

"Simply that I will cast my vote with disinformation because its effects are easier to predict and control. For about a month the Russians regularly spotted crescent-shaped lights in the sky, and they were thrilled at the thought they were seeing extraterrestrial spaceships, which discredited the story even more."

"That's not good enough in this case, Greg. That was way back in the sixties. 1967, if my memory serves. We don't have the same luxuries as the KGB. People, civilians, must have seen this object. Hell, it was so bright, they must've seen it from the Las Vegas strip! And they must have seen our plasma beam. Even the radar people will leak the story eventually. How was I supposed to know this object was gonna get so close? We don't have time to plant sophisticated rumors."

"Why can't we simply deny the whole business one more time?" Joe Wilkinson volunteered. "Plant an article in *Aviation Week*. You know, radar malfunctions and the like. The skeptics will spread it all over the media. Makes them look smart."

"I told you, denial is not a feasible option," the general said. "Neither is obfuscation. That won't work any more. Today in America you've got too many people who are aware of this phenomenon. Before you know it they'll be picketing the White House and calling for Congressional hearings. They'll be flaming your skeptics and each other all over the Internet. No, it's no use denying that UFOs exist."

"Then I'd rather take the openness route. Come out and tell the whole story," said Greg Parker.

"Too unpredictable," countered Wilkinson. "You don't know how people would react."

"I vote for disinformation. The ideal solution is to work through the believers themselves. Let them provide their own cover-up," proposed Beverly.

Wilkinson leaned forward: "Like the British Ministry of Defence did with crop circles, disguising their infrared laser tests as a New-Age phenomenon?"

The general nodded: "I like that a lot better than denial. If we're going to use disinformation effectively, we have to leak out part of the truth."

"Which version of the truth do you want us to leak out, sir?" asked Beverly ironically.

Bushnell smiled.

"Ours, naturally. You don't expect me to release the film of this baby playing with my jets, do you? We need to study it in peace, even as we accelerate the last phase of our plan."

"We'll have to build a whole new legend, sir. There are no cornfields in Nevada where we could draw magical symbols . . ."

". . . and Las Vegas is a long way from Stonehenge . . ." Wilkinson chimed in. "We're not in mystical territory here."

The general leaned back in his chair.

"All right, all right. The decision is made. So you'll need to find an advocate. A thoroughly articulate voice. A credible witness who can be our interface."

"How do we do that, sir?" asked the astonished Parker. "Who are we gonna pick?"

"I don't care," said the general, abruptly getting up. "Just find me somebody who can fit into the plan. And do it fast, whatever you do."

CHAPTER THREE

The house was a sturdy structure surrounded by a Gothic atmosphere, a somber old brownstone on the Upper East Side. The fast-moving dark clouds that blanketed the tops of nearby highrises cast an even gloomier spell over the stone steps, the ironworks, the dripping gutters.

Inside the house, in the second-floor flat, a six-foot man was sleeping heavily on a five-foot couch. When the television playing in the next room awakened him, he rolled over, bones creaking, and tried to clear his head enough to identify the sounds coming through the wall. He heard rap music turn into a news story about riots in Mexico, then comments by President Henry Templeton. This changed into the Three Stooges, which became a politician from the Christian Right discussing the New World Order, which became an exercise show.

In the next room Sara Keller, a doe-eyed nine-year-old, was sitting cross-legged, remote control in hand, on the edge of the recently slept-in bed. The TV screen reflected off her face; now an athlete was imploring his viewers to exercise along with him.

She hit the button again and the image changed. Dr. Borodine's voice boomed through the apartment, irritated and supercilious.

"And people say, *Doctor Borodine, how come the aliens don't just land on the White House lawn?* And I say, *What a stupid, ignorant question! Why should they land in President Templeton's front yard, when he doesn't even believe in them?*"

The little girl made a face and changed back to the exercise show. She had started hopping along to the music led by a muscleman and his Amazonian sidekick by the time her father opened the door behind her and watched with a tired smile.

Peter Keller hugged his daughter. At thirty-eight, he was still lean and strong, with a sardonic smile on his face. His hair was a little long for this decade—he looked like a sixties throwback.

"What's the matter, Pumpkin, no cartoons on Sunday morning?"

Sara resumed her jumping.

"There's plenty of cartoons, Daddy, but I need the exercise. Gérard says I'm three pounds over my ideal weight, you know. I should watch my diet."

Talking about Gérard was always a sore point. Peter changed the subject: "Thanks for reminding me," he said, trying to sound cheerful. "It's breakfast time."

* * *

They sat at the small table in the tiny kitchen. The cluttered look of the place made it obvious that her father had sacrificed space in order to live in Manhattan. He placed boxes of Sugar Frosted Flakes and Raisin Bran before her.

Sara frowned.

"Dad, this is fifty-percent sugar," the girl complained.

"What? You've always loved Frosted Flakes!"

"Not any more, Daddy. Don't you have any Natural Oat Bran? Or Unsalted Wheat and Fiber?"

"Jesus, Squid, that stuff tastes like pencil shavings," Keller replied, trying to draw a smile from her. But the little girl was as stern as a schoolmistress.

"You should start eating it too. It stops you from getting cancer."

"That's bull, honey. Madison Avenue hype."

"But Gérard says—," Sara began. He interrupted her:

"Gérard says ... Gérard also says baseball's boring and Tai Chi's exciting. Yeah, about as exciting as combing my hair."

She giggled. "You just don't like Gérard because he's Mom's new boyfriend."

He shot her a surprised look, eyebrows arched, but she only giggled even more, so he opened the box and filled her bowl, acting with all the fatherly authority he was able to muster.

"Eat the Raisin Bran, then."

She shook her head.

"There's sugar on the raisins," she pointed out reproachfully.

* * *

"No one in his right mind would be caught out in this weather," thought Peter as he wiped the dripping water from his face.

The corner market was run by a black man in his late fifties who broke into a twisted grin when he saw Keller, sopping wet, standing at the counter with a bottle of aspirin, a carton of seaweed juice and a box of "Unsalted, 100% Organic Oat Bran."

"Feeling healthy today are we, Pete?" the guy asked cheerfully.

Keller shot him a look that said, "Very funny," grabbed the paper bag and psyched himself up for the dash outside. He opened the door and ran straight out into a driving wave of freezing rain. Behind him he could still hear the store owner laughing heartily.

Peter only noticed the woman after he was caught in the flash from her camera. She was big and fat, and there was something odd about her feet—something he would often think about later, when he woke up for no reason in the middle of the night and had trouble finding sleep again. She was wearing fur-topped boots and a flowery dress that was longer than her coat. Her clothes were cheap, the kind she might have picked up at a Goodwill store, but there was nothing cheap about her camera, a zoom-lens Nikon with a sleek flash unit. A second camera dangled from her left shoulder. Peter was stunned.

"Why the hell are you taking my picture?" he asked her before crossing the street.

"Just trying out my new film," she replied, unshaken.

"This town is getting crazier every day," Peter thought, rushing home through the rain.

* * *

The sun came out briefly, unconvincingly, in the afternoon when Peter and Sara drove out to the house in Westchester County, the house where they had all lived together for years—until the divorce, that is. Kathleen stepped onto her rain-soaked front porch to meet them, and Peter had to admit that she was still striking, even with her dark hair pulled back in a simple pony tail.

Sara ran up to hug her. Peter studiously ignored the scene. He walked around the side of the house, toward the back yard.

Little Sara looked concerned. "I'm worried about him, Mom. He's not taking good care of himself."

"What else is new?" asked her mother.

But Peter was kneeling in the yard, stroking his golden retriever, dog and master clearly happy to see one another again. In the background, Gérard was slowly practicing complicated Tai Chi moves. The edges of his big straw hat flapped around his face while he sliced through invisible enemies with his extended hands. He was a wiry fellow, with a little mustache and a golden earring in his left ear.

"You should wear something on your head, Peter," Gérard recommended with a serious frown. "The ultraviolet rays are especially bad after a rain. You can get skin cancer, you know."

"Damn, now why didn't I think of that? Thanks so much," answered Peter, with a big, fake grin.

"It's never too late to become aware of the environment, even for you. The way the industrial-military complex is depleting the ozone layer, we're all gonna be in trouble soon. By the way, how's the novel coming?"

Peter made a move to get away.

"It's coming," he said over his shoulder.

"Two and a half years now, eh? Must be a real epic."

Kathleen spotted Peter as he was heading back toward his parked car. He seemed cheerful, practically whistling, but she knew it was all a front. He was dying inside.

"So that's it?" she asked. "You only say hi to your dog?"

"No time for anything else; gotta go buy an aluminum hat," replied Peter with urgency in his voice. "The cosmic rays are everywhere, you know," he added as he got into his car. "Drives everybody crazy. Fat women are taking my picture in the rain. Oh, and do tell Jules—"

"—Gérard."

"—whatever, that I'm not letting him turn my daughter into a sprout-eating aerobics zombie. Got that, babe?"

"Oh, Christ, Peter, grow up."

"You need a new line, Kat," he said, revving his engine. "I've been hearing that one for years."

"It still fits," she muttered to herself as she watched him go, with more than a trace of sadness.

CHAPTER FOUR

It was raining again the next morning when Keller stepped off the downtown bus. He ran along with the crowd, and took the elevator to the thirtieth floor of the Manhattan building where he worked. A few minutes later, his raincoat on his arm, he stepped as calmly as he could manage into the offices of *Inside Story,* wearing dark glasses and fighting a heavy-duty hangover.

A flamboyant sign on the door proclaimed: SIMON PRODUCTIONS. Strands of circus music filled the air, and the receptionist smiled warmly at Peter, but none of that could lift him out of his foul mood.

"Man, you look like death," commented Chad Burke, one of the show's latest recruits, as soon as he saw Keller. His Australian good looks and his deep tan contrasted with Peter's drawn face and tired spirits.

"Come on," he replied, trying to sound cheerful, "even death looks better than me, especially on Monday morning. Where's my friend Tommy?"

Chad didn't bother to answer, simply pointing at one of the darkened editing rooms that lined up along one of the walls.

Tommy Archer, a black man in his early thirties who had a dry sense of humor and a knack for turning the most innocuous story into an epic of historic proportions, was already at work, reviewing a tape of the show's next installment. Peter stopped in to watch it with him.

"I thought we'd begin with my eerie segment about Fred Barrington, who is basically your everyday, quiet, church-going executive turned serial killer," Tommy said.

The camera followed the man as he walked quietly through the streets of his town, over a text narrated by Chad Burke with the arrogance of a former anchorman.

"Man, this guy is pompous!" commented Peter.

"Wait till you see the rest of it!" answered Tommy, hitting the Forward key.

In the next segment, a man who called himself Nik Tok postured and raved about what he called "the golden opportunities of infinite wealth," before a background of palm trees and bikini-clad beauties. He was a promoter from Asia who claimed to have arrived with the boat people and made an instant fortune in telemarketing. Dressed in an impeccable white suit and white shoes, he was hawking his expensive seminars and extolling the virtues of quick money, smiling with wall-to-wall teeth.

"That guy looks like a real jerk," Peter remarked, "almost as bad as the serial killer. Why are we giving them so much air time?"

"You ought to ask Ben that question," answered Thomas. "I don't make the policy around here." He shrugged, adding, "Cheer up, buddy. We are blessed with a tabloid audience that's easily impressed by quick social success."

In the next room their fellow producer Holly Sukuda looked up from her screen as a phone call came in, and announced:

"Peter, the boss wants to see you."

"Lucky you," joked Tommy, "maybe he'll give you another assignment like the one we did last week, about the mystery of Marilyn Monroe's face appearing on an old refrigerator door in Michigan . . ."

"Or the old woman in San Diego who died of spontaneous combustion after eating too many burritos," added Holly as Peter walked off heavily towards the corner office, pretending not to hear them.

* * *

Keller found Ben Simon watching the big monitor in his plush surroundings, both feet propped up on his desk. A short, barrel-chested, cigar-smoking, sharp-minded supervising producer who affected utter indifference towards other people's feelings, Ben Simon had nonetheless succumbed to the vanity of surrounding himself with personal photographs. Peter quickly scanned the pictures on the walls: Ben with family, Ben shaking hands with Hillary Clinton back in 1996, Ben with movie stars, and numerous framed articles about *The Inside Story.* The main subject in all the photographs was none other than Ben himself.

"What's going on?" asked Peter, attempting to sound cheerful and bright.

Ben Simon made a sort of welcoming gesture, vaguely moving a hand up in the air, but he never took his eyes away from his TV set, where the picture showed a round-faced preacher in his mid-sixties who ranted and raved about the imminent end of the world, which he kept calling "The Eschaton."

"Can you believe this guy? What a gold mine. What a racket!" Simon cried, slapping the top of his desk and squirming with enthusiasm.

"Who's he, Ben?" asked Keller flatly.

"Helmut Borodine. No, forgive me: *Doctor* Helmut Borodine, the fearless leader of the BGS—The Brotherhood of Galactic Science, no less. Based in sunny Tampa. Sit down and listen. You might learn something."

The man was lecturing from a rocking chair in front of a huge plexiglass board with the words *The Brotherhood! They're Coming to Earth!* and *Show your Faith, Pledge Now!* written all over it. On the wall above him were seven clocks, each showing a different time. They were labeled: New York, Los Angeles, Tulsa, London, Paris, Tokyo and Jupiter. A banner above the clocks read: *Only One More Year! Prepare Thyself for the Eschaton!*

Borodine, who was wearing a light-colored three-piece suit, was a sharp-eyed, white-haired man with a neatly trimmed beard. Sometimes he sounded like an evangelist, at other times more like a mad scientist, quoting interminable statistics no one could ever verify, or scribbling meaningless formulas on the board, and occasionally stopping in the middle of a speech to accept a glass of what he called his power drink from the hands of one of his female assistants.

Keller sighed and turned as if to leave, but Ben imperiously motioned for him to stay. He turned up the volume, and Borodine's raspy, growling voice filled the room.

"The Space Brothers are coming," he was practically screaming at the audience, "and that is why we must urgently build the Solid Gold Landing Strip: a twenty-four carat welcome mat for Our Brothers, the Aliens From the Stars. When this task is complete, they will magnificently touch down, put the Earth's pollution under control, and save us from ourselves!"

Ben visually fast-forwarded the tape a little, then let it run again. Peter saw the Brotherhood's eager disciples, row upon row of the ever-faithful Voices of Science, busy volunteers taking cash pledges on the phone for the good doctor as an 800 number flashed on the screen.

Ben Simon raised a finger: "Pay attention," he told Peter, "this is the good part."

Borodine was reading through a thick stack of donation slips, tossing each one to the ground after reading it out loud.

"Phoenix, 500 dollars ... Memphis, 84 dollars ... Dulce, New Mexico, *Dr. Borodine, Sir—Thank God you have the courage to tell the truth!*—one thousand dollars."

There was brief applause from the studio audience. He looked up, scowling. "What's the total, Brother Ralph?"

Ralph Augustine was a suave, obsequious Hawaiian in his early forties who evidently served as Borodine's New-Age accountant and sidekick. Dressed in a gaudy green suit that did nothing to hide his rotund figure, he stood at a little podium, working a calculator.

"Two hundred and forty-three thousand dollars and twenty-seven cents," he announced triumphantly.

But when the camera moved in for a tight close-up of Dr. Borodine, he was looking directly at the home viewer with fire in his eyes.

"*Not good enough!* We're still *far short* of our goal. Is that how you expect to overcome the Government Cover-up? With a lousy two hundred thousand bucks? If each and every one of those phones doesn't light up in the next ten seconds, I'll know you're still not ready to evolve along with Galactic Science! You're not ready for the spiritual transformation of the New Age! Do you want the Eschaton to find you unprepared? Or will you join us in glory to Witness the Tribulation, like the Good Book says?"

His voice turned into a growl:

"Get on the telephone! Brother Ralph, sound the charge! Let'm know we mean business!"

A bugle call blared throughout the BGS studio. When the final note faded, Borodine raised his arm into the air like an orchestra director, and the Voices of Science sang out in unison:

> One, two, three, four,
> The Aliens deserve much more!
> Five, six, seven, eight,
> Make your pledge, and don't be late!

The studio camera zoomed in and out wildly. The 800 number filled the screen again. Borodine rubbed his temples as though fatigued or deeply annoyed. He ripped off his lapel mike and abruptly walked off the set. The camera stayed focused on the empty chair, and the baritone voice of Ralph Augustine was heard:

"Doctor Helmut Borodine *will not return* until we have met tonight's pledge goal. Doctor Borodine has dared to share the truth with you, the truth about the Millennium, the truth about the Eschaton. Won't you give him a testimonial to his teachings? *Invest in The New Age!* The time to give is now! Before it is too late!"

The image of the empty chair froze as Ben hit the pause button. Keller laughed, shaking his head.

"They collect twenty-four hours a day," said Ben, pulling out a handkerchief and wiping beads of sweat off his forehead. "They hawk New-Age shit, they show flying saucer movies up the ying-yang. But the bottom line is money, what else? Wow! This guy really knows what he's doing."

He lit up another cigar, pushed a pile of papers to a corner of his desk and turned to face Keller. Suddenly he was all business.

"That's your new assignment, Pete. The Brotherhood. The Ufological Community."

He pronounced it "OOO-Fo-logical," as in "oodles."

Peter stopped laughing.

"Jesus, Ben, you've never made me do UFO shit. This is an all-time low."

Ben jumped to his feet, suddenly furious.

"You always fight me when it's something good," he screamed. "Remember the dwarf-tossing piece? You bitched to high hell, but it was one of our most popular segments ever!"

"Schlock," Pete stated again forcefully. "Next, you'll be showing fake alien autopsies."

"*Schlock sells,* don't you get it? What the hell is the matter with you?"

Ben drew a big puff of smoke from his cigar, sat down, leaned back and went on in a more amicable voice:

"Anyway, this ain't about 'oo-foz,' Peter, it's about all the morons who buy this kind of shit, you get the point?"

He got up, pointing a stern finger at Keller's chest.

"We've wasted enough time. Get to work. Tommy's already busy checking out the cults. I want you guys rolling first thing in the morning."

* * *

Keller returned from Ben's office with a slow, exhausted look, which didn't lift from his face when he found Thomas Archer already at his desk, looking at a computer display, typing away.

"So it's fruitcake city again, huh, pal?" He sighed as he sat down heavily across the big table from his partner.

Thomas didn't answer. He was busy printing out the list he had compiled on the screen.

"Thank God for Internet! Every crackpot has got a home page these days," he mumbled to himself. "In the old days this'd have taken hours."

"So where do we start?" asked Peter. Thomas pulled the list from the printer.

"Let's see ... we got The Clan Of Koltar, the BGS, the Institute for UFO Analysis, the Antarians, the Lemurians ..."

"The Lemurians, that sounds good. Where they from?" Peter insisted.

"Uh ... Bismark, North Dakota," said Thomas dubiously.

"Scratch them. How 'bout the Anteaters?"

"You mean the Antarians?"

"Whatever. Where's their home base?"

"Antares, I think. It's a star in the Scorpion constellation. My wife is nuts about astrology."

"Don't play games with me. Where are they *on this planet?*"

Thomas checked his list . . . and smiled. "Key West," he said. "We can be there tomorrow."

A look of complicity passed between them. They had the same thought: At least it was a chance to get out of the rain.

"And the BGS is in Tampa. We'll do both in one trip. Save on expenses."

Ben was sure to like that.

* * *

When Peter and Thomas arrived from New York on Tuesday morning, the first thing they saw was a big sign that read *Welcome to Key West*. They wheeled their equipment out of the terminal under the hot sun and loaded it into a taxi.

They found the Antarian headquarters without any problem. It was a former Chevrolet showroom, skillfully converted into a New Age shrine. Outside, a huge mural showed an Egyptian scene with pyramids, a row of statues, a pink Sphinx and a flying saucer in a deep blue sky, beaming down towards the parking lot. Inside, everything was draped in a velvety darkness relieved by a few colored spotlights.

Madame Venus, the spiritual leader of the Antarian Society, greeted them in person. She was wearing a long gown spangled with stars, planets and runaway comets. She sported a beehive hairdo and spoke with a quivering voice.

"We built the Museum of Antares with donations from a few enlightened Earthfolk who are spiritually elevated," said Madame Venus, batting her false eyelashes with fierce energy. "Each love brick, each block of love wood, was cosmically blessed by Voleboar, King of Saturn."

Madame Venus glided about the little museum. Peter saw other robed figures milling in the background. The place was cluttered with star charts, pyramids, plastic space flowers, a rotating radar antenna and, of course, bronzed UFOs of various sizes and shapes. Thomas began filming her while an amused Keller looked on.

Madame Venus halted beneath a golden UFO dangling from the ceiling. She used a silver wand with a star on the end to point to a life-size statue of a snarling, hairy giant.

"As you know, Bigfoot comes from the Andromeda galaxy. It's all explained on our CD-ROM. He tried to land his saucer at Holloman Air Force Base but the military shot at him, forcing him to flee into the redwood forests where he now makes his home."

"Redwood forests? Around Holloman? Isn't that base in the middle of New Mexico?" asked Peter incredulously.

"And you thought this wasn't gonna be scientific," Thomas whispered into his ear.

"Shows you how little I know about the New Age," Keller sighed. Suddenly, for no reason, he started thinking of Gérard.

"Remind me to buy a hat when we get out," he told his partner.

CHAPTER FIVE

The F-16 fighter, a beautiful, sleek aircraft, was positioned on the runway. The cockpit was empty and the engine was on, whistling quietly. Captain James Hall stepped out onto the tarmac, flight helmet in hand, looking worn out and a little unsteady from his long interrogation, but happy to see his airplane again. The bearded man was by his side as he walked to the fighter.

"I don't know who you people are . . . I don't know, and at this point I don't care. But I assure you that, soon as I get back to Edwards, I'm filing a full report. I'm gonna bust your ass for treating an Air Force officer the way you've treated me."

The warm wind was bringing whiffs of wild smells from the desert.

"You do whatever you have to do, Jim," the bearded man answered sadly, quietly. "Sure you don't want to sign that document?"

Hall looked at him with fire in his eyes, wondering if he was joking, but the man's face showed no expression. They'd reached the airplane. Hall fastened his helmet and climbed up into the cockpit.

"Happy landing," the bearded man said simply, backing away from the plane, watching it turn to the main runway and rise rapidly into the sky.

* * *

Up above the mountains, Captain Hall spotted trouble as soon as he pulled back on the throttle. The exhaust gas and oil temperature gauges were rapidly rising into the red. Fuel pressure was plummeting. Red lights went on, all across the instrument panel.

Suddenly, James Hall knew why things had gone wrong. And he knew what would take place in the next few seconds. Before he could activate the ejection seat.

On the runway below, the Beard's lone silhouette watched the ball of fire over the distant horizon.

"You should have signed, Captain," Greg Parker mumbled to himself, scratching his head. "After all, it was just a lousy piece of paper."

* * *

"So we're back in the mind-control business, General," noted Parker as he sat down at the long oak table where most of the planning sessions usually took place.

"It didn't work too well with that pilot," commented Beverly Bernard. "The one we terminated."

"Mind control is a question of time," said General Bushnell, "and in this particular case time, as a commodity, was in very short supply."

He grunted, "Need I remind you once again that we must move quickly? What happened in Nevada was a historical opportunity, but it wasn't part of the original plan. By the way, how well did you secure this facility?"

Greg Parker looked around them.

During the seventies and eighties Alintel had constructed high-tech secure vaults near all major metropolitan areas, working under cover of the "Continuation of Government" program of the Federal Emergency Management Agency, whose annual budget ran into cool hundreds of millions. This particular vault happened to be in the basement of a nondescript office building on the Washington Beltway, in plain view of thousands of businessmen and government bureaucrats who drove by every day on their way to Reston or Dulles airport.

A big neon sign on top of the building read BOEING, but this particular wing of the sprawling structure was managed by McGraw-Hill Publishers, and the basement was leased to Science Applications, Inc., so that anyone trying to reach that particular room had to be cleared for three different projects, sign on three sets of books and receive three different badges. And none of the three legitimate companies had any notion of what actually went on between the concrete, copper-lined walls of the bunker downstairs. More importantly, since the vault was located on private property rather than inside a government facility, nosy Congressional staffers would have to deploy extraordinary legal efforts to gain access to it, assuming they ever heard of Alintel in the first place. All they would ever find inside the room would be a map of the world, a video-conferencing transmit-and-receive system equipped with document cameras, and a bank of phones crammed with standard encryption chips.

* * *

"Do you understand the scope of the mission?" asked General Bushnell, resting his eyes in sequence on every member of his staff, stopping on the youngest recruit, the bearded man. "Dr. Parker, you're the newest addition to this team. Why don't you tell us how you would approach it?"

"I have read the briefing materials, sir, on the way back from Area 51," he answered, proud of having General Bushnell as a mentor. "I don't have as complete a view of the problem as you do, so I think of the assignment primarily as an experiment in the psychodynamics of absurdity."

"That's a good place to start," the general conceded, "although others would tend to look at this purely in terms of conflict. After all, we are trained members of the military."

"I find it hard to think in terms of conflict where extraterrestrial forces are involved, sir. Especially given the nature of the cards we're holding. And the secrecy. We are not just another part of the government, after all. We constitute a cryptocracy . . ."

". . . yet you correctly offered the view that we were engaged in a social experiment," Beverly broke in. "Don't forget that. We bypass the White House."

"Precisely, we are a human experiment that deals with nonhuman forces."

"Doesn't Alintel routinely brief every new president on its mission, shortly after he is inaugurated?" asked Greg Parker.

"Depends," replied Bushnell. "What little he absolutely needs to know we've buried in a vague presentation on prospective weapons technology that he gets from the Air Force. We do have to tell him something, if only so he can keep up the dialogue with the Russians. But to be completely frank with you, the briefing is a snow job."

"Does it talk about where we found all our little toys?" Parker went on.

"Heck no! Why should we reveal how close the limits of our ignorance are to the threshold of our ambitions?"

Parker smiled but he insisted: "Templeton isn't aware of us then?"

"He certainly hasn't been fully briefed," answered the general. Clinton wasn't. Reagan and Bush did know about Alintel, and we had handlers for their underlings, but the Executive has been in the dark since then."

"What about earlier presidents?"

"Truman knew. Back then we had just recovered our first few bits of alien hardware. We became arrogant. We thought the solution was in sight."

"Eisenhower?"

"Of course."

"Anyone else?"

"There was no need after that," said Bushnell with a gesture that suggested helplessness, failure. "We knew better. The research went nowhere. Under those circumstances the Administration is of no concern to us, no concern whatsoever," he added with emphasis. "We have to negotiate our human identity with other beings of unknown scope and function. This is not just another episode of traditional policy."

"We are boundary spanners," commented Beverly.

"All that may change," said Wilkinson. "From now on the Fastwalker is the basis of our transactions."

Parker grunted. "It's much more than that. In human terms, in terms of the value pattern of dominant elites on this planet, the Fastwalker represents a violation of expectations, pure and simple. A violation as drastic as any event in the history of humanity."

"I disagree," said the general. The others stared at him.

"It's only unusual in terms of the level of the technical challenge," he went on. "It's certainly not new in terms of power relationships. Think of the Manhattan project. That, too, was unprecedented in history."

"He who has the power can win over less fortunate players," pointed out Beverly in a quiet, measured voice.

"That's hardly the point," replied the general. "The bright folks who wrote the *Iron Mountain Report* as a joke, arguing for the creation of a fake external menace as an economic substitute for the benefits of warfare, made the same mistake as you."

"What is the point, then?" asked Beverly.

"The point is, he who has the power can do far more than win, he can define the game being played."

"Which brings us back to mind-control," Beverly said quickly, leaning forward in her seat. "We're wasting valuable time, when we should be moving the plan ahead. With all due respect, sir, I am tired of hearing about such subtleties, when we have so many convenient weapons in our pharmacopoeia, to bend the public to our goals. That is, assuming we're serious about defining the game, as you said."

"Why not use both? Physical and psychological means?" asked Parker.

"Dr. Bernard is right, of course," commented Bushnell, unrustled by the heat of the debate. "There is no better tool, in a strategy of the absurd, than acting directly on behavior. How much do you know about all that, Greg?"

Parker thought it over. "I am reasonably familiar with the MK-ULTRA material, the stuff that came out in the seventies, but I haven't read all the classified data."

"I want you to read it, then. First thing tomorrow morning, go over to Langley under your ARPA cover, tell them whatever you want, go downstairs and read the damn file. Don't start in the seventies, though. Start with BLUEBIRD, April, 1950, when Director Hillenkoetter at CIA authorized the use of unvouchered funds to pay for the work on behavior control."

"Was that the same Hillenkoetter who was the first president of NICAP, the civilian UFO group that kept complaining about a cover-up?" asked Greg.

"Are you shocked?" Beverly laughed. "People only think of Major Keyhoe when they think of the National Investigations Committee on Aerial Phenomena. They don't look behind the façade. Yeah, Donald Keyhoe was a true believer. He was always pestering the Air Force, demanding Congressional hearings about flying saucers. But his own board read like a *Who's Who* in strategic deception."

"Can you give me some names, at least?" asked Greg in astonishment. He only knew Keyhoe from the old newsreels, as a rather intense and obviously sincere man who had written some of the most influential books on flying saucers in the fifties. More than anyone else, he had promoted the view that the saucers were nuts-and-bolts craft that came from outer space.

"Look up the careers of the other board members, when you're over at Langley tomorrow. Three of them were career Intelligence officers with a specialty in psychological warfare."

"What are you saying? That NICAP was a collection channel for the CIA?"

"It certainly served that function, as every UFO group has done since then, knowingly or unknowingly. But it also served to foster the notion that UFOs were extraterrestrial. A very useful bit of propaganda, for our purposes."

"So much has changed," said Wilkinson with a smile. "In those days the saucer believers went around looking for physical proof, photographs, hard evidence, physical saucers from space. Now it's all hypnosis. It's the psychological stuff that's important for them. Look at all the books about missing time, it's all Suggestion and Company."

"So much has changed, yet so little has changed," corrected Beverly. "The CIA Office of Security began contemplating the operational use of hypnosis way back in the forties. Doesn't that make you wonder? A fellow named Sheffield Edwards formed interrogation teams that used hypnosis. He hired psychiatrists and polygraph experts, and trained them."

"I remember reading Edwards' stuff," said Wilkinson. "Whatever happened to him?"

"A decade later he was handling joint CIA-Mafia operations. I lost track of him afterwards."

"I take it the research didn't stop there?" Greg said, fascinated.

"Hell, no! BLUEBIRD turned into ARTICHOKE, whose goal was to develop technology to wipe the human mind totally clean. Between 1950 and 1952, responsibility for mind control at CIA went from the Office of Security to the Scientific Intelligence Unit and back to Security again. Good luck tracking down the files, Greg," Beverly concluded.

"Could they do it? I mean, wipe the brain clean?" Parker asked.

"Not very well. They did turn people into vegetables, so to that extent they were successful."

Beverly Bernard chuckled and went on: "You might say that the Artichoke lived up to its name!"

"And that led to MK-ULTRA, then, in the seventies," Parker insisted.

"No, no, you still have the chronology all wrong. MK-ULTRA had been approved by Allen Dulles way before that—on April 13, 1953, to be exact."

"You've got a good memory for figures and dates."

"Memory can be trained. You shouldn't be in this business if you can't remember the key facts. Dulles had exempted the program from all normal CIA financial controls, which was very convenient for the Technical Services Staff. Think of it: No signed contracts to be found by nosy investigators, no written agreements to embarrass the major universities that participated. They signed up Cornell, McGill in Canada, many others."

"And I wouldn't simply search under the codename MK-ULTRA, if I were you," Wilkinson added helpfully. "You should also look at a bunch of others, like MK-DELTA, which governed the use of our chemical and biological warfare products. It was the actual operational arm, designed 'to investigate whether and how it was possible to modify an individual's behavior by covert means.' That's an exact quote from Sid Gottlieb, who reluctantly testified about it before Congress a few years later."

"And the public never saw any of those products?" asked Greg.

"Where have you been, kid? Didn't they tell you about LSD, over at Berkeley? We derived LSD from ergot. The French and the Swiss had the idea first. We were looking for a means of generating differential amnesia. With ergot, one of the first symptoms is extreme paranoia and occasional fainting. As a biological agent, it's very convenient. It mimics a hundred and forty-four different diseases. It also produces a form of AIDS that doesn't give an HIV-positive test response."

"And all that work was done without any of the money ever being traced?"

"A small part of it was traced by Congress," Bushnell told them. "But much of the dough was funneled through the *Society for the Investigation of Human Ecology,* a name which had a nice environmental ring to it. Not that anybody suspected anything, of course. The young generation had already adopted LSD as a symbol of anti-establishment rebellion, which suited us just fine, of course ... By then they were too stoned and too happy to care where the stuff came from. Just like now, with the psychedelic mushrooms from South America."

"Until the scandal at McGill, with Ewen Cameron," remarked Wilkinson.

"That was only a minor incident, although an unfortunate one, to be sure," said General Bushnell with a wink. "Professor Cameron was careless at times in what he fondly called his 'depatterning' work. But what an interesting fellow! Do you know that at the end of World War II he came up with the suggestion that every surviving German over the age of twelve should receive a short course of electroshock treatment, to burn out any remaining vestige of Nazism?"

The group broke into bitter laughter.

"Too bad he wasn't around when we needed him, when we ran the People's Temple operation in San Francisco," remarked Beverly. "By then there were lawsuits all over the place, and MK-ULTRA was severely compromised."

"Is that when the work stopped?" Greg Parker went on. "After the Cameron exposé?"

"You are more naive than I thought, my young friend," Bushnell said. "Why should the work have stopped? The cover was compromised but it certainly wasn't destroyed. Want proof? Go out on the streets of New York or Chicago tomorrow, and try to find ten people who remember what MK-ULTRA was all about. Nobody gives a damn. Even after all the fuss on Capitol Hill, and the public confessions, and the books that came out."

"What happened to the project, then?"

"It was redirected into ESP research. That wasn't a new direction either: As early as April 9, 1952 you will find a memo that suggested looking at the influence of drugs on human paranormal abilities."

"Was it ever acted on?" asked Greg in astonishment.

"Of course! Among other discoveries they found that barbiturates, notably sodium amytal, decreased telepathic performance. In fact, they were so successful that most of our brainwashing work was later channeled under parapsychology."

"And here we are," concluded Wilkinson, "except that Alintel has left the old CIA paradigm far behind. And now, to top it off, we are the proud owners of a flying saucer."

"That's quite a story," said Greg.

"Welcome to Reality Zero, kid," said General Bushnell. "Stick with me, you'll learn something." He patted Greg on the shoulder. "And now let me give you your assignments for the next couple of days."

CHAPTER SIX

In person, Brother Ralph was far less intimidating than the fast-talking huckster Peter and Tommy had observed on the screen. The jovial, burly Hawaiian was now dressed in polyester slacks and a white sportscoat over a bright silk shirt. He was waiting for the team from *Inside Story,* standing before an attractive building with wall-to-wall tinted glass and a battery of huge satellite dishes on its roof.

"Welcome, friends," he told the reporters in his distinctive mellow baritone. "Behold the home of Galactic Science."

They followed him down a vast linoleum-floored hallway, passing elaborate paintings of Dr. Helmut Borodine in various poses, ranging from thoughtful (leaning against a globe in his library) to proud (supervising the work on the Headquarters building). In some of the portraits a subtle halo emanated from the good doctor's head. Each expensive frame was complete with a Helmut Borodine quotation in gold letters.

Nobody had to tell Tom Archer it was time to start filming.

* * *

They followed Brother Ralph on a quick tour of the BGS headquarters, which took them through a series of large exhibits depicting various facets of what he called "New-Age Technology," including models of Ion-Gravity Pulse Generators labeled *The Secrets of UFO Propulsion.*

Next came a room jammed floor-to-ceiling with electronic equipment, where a dozen young men sat at consoles, wearing headphones.

"This is our own version of SETI, the Search for Extraterrestrial Intelligence," Ralph said, pointing at the banks of digital signal analyzers that lined the walls. "Our scientists are listening for further contact from the aliens," said Brother Ralph, intently staring at the camera lens. "So far, the spacemen have only spoken to Doctor Borodine, and always telepathically. But we do have recordings of highly significant signals."

"What do they say?" asked Thomas.

"They're yet to be deciphered," Ralph answered. "All that scientific work takes money, you know. Perhaps people will send us some contributions, after they see your show."

They passed a copper-sheathed room where members of Borodine's inner circle, wearing sterile white suits, were seated in a circle around a shiny device made up of multiple spiral-shaped plates. They were extending their hands, palms-up, towards the instrument, and they chanted softly.

"What's that, a giant slinky?" Peter asked in astonishment.

"Don't disturb them. They're charging a bio-cosmic battery," whispered Brother Ralph respectfully. "One day, all the power needs of an entire planet will be met by one such device."

"How does it work?"

"It draws out the power of the vacuum. Dr. Borodine says it uses the Casimir Force."

The building also contained a daycare center, where the children of the faithful could play with big, lovable, furry E.T. dolls while others whirled around in a flying saucer carousel.

"Yes, my friends," Brother Ralph went on. "I owe it all to Professor Borodine. Before I joined the Brotherhood I was selling rugs in Los Angeles."

Walking behind him along the corridors, Keller and Thomas had trouble keeping a straight face. Borodine's portraits gave way to oil paintings depicting tall, blond, angelic robed figures standing next to landed saucers, waving at the viewer in friendly fashion. Each canvas had a homey religious feel.

"What made you decide to come here?" asked Peter.

Ralph lowered his voice respectfully: "When I saw Doctor on TV and heard the Word of the Space Brothers, it changed my whole way of thinking. I'm no longer afraid of the Eschaton. I'm ready for the New Age."

* * *

Brother Ralph led the visitors through a side door and into the wings of a TV set where a live show was in progress. A studio audience of some fifty people, together with the Voices of Science in purple robes, looked on as a pacing, scowling Borodine railed at his listeners. Behind him, the words *Ro-Man, Mayan Vortex, Zero-Point of the Vacuum, Negative Energy, You're Next! Skeptics, Overlords, Missing Links* and *Watch the Skies!* were scribbled on the blackboard. The professor sounded even more upset than usual.

"I'm getting good and goddamn mad at those bonehead skeptics—and their allegedly scientific reasons why UFOs supposedly can't exist!"

Borodine spoke angrily, straight into the camera.

"Well, Mr. and Mrs. Skeptic, I'm telling you here and now that the Saucers are real! The masters of these spacecraft have shared their deepest secrets with me. *I know the truth!* And the Truth shall make you free. Well, wind 'em up, get 'em going !" He yelled at someone off-stage.

Thomas and Peter were puzzled to hear growling mechanical noises in the background, with sounds of whirring, motors running, whistling and buzzing.

"My Voices of Science know what these skeptics are ... Well, don't you?" he added, growling at them.

"Yes, sir!" They screamed almost in unison.

"Those self-appointed critics spew forth skeptical nonsense like a bunch of monkeys, don't they?"

"Yes, sir!"

"Well, show them what I mean!"

The camera awkwardly panned to reveal two dozen battery-operated monkeys, making their solemn entrance on stage. Dressed in tiny white lab coats, they uttered excited chimp sounds. Some banged cymbals together, others did somersaults, played xylophones or hit themselves over the head with tiny hammers.

"Let's give these idiots what they deserve," the professor ordered.

The Brotherhood band whipped the faithful into a frenzy, and an enraged Dr. Borodine seized a plastic bat and started bashing the monkeys on the head, scattering bits of metal all over the studio. Finally satisfied, Borodine tossed the bat aside, ripped off his lapel mike and stormed off the set to the raptured applause of his followers.

Brother Ralph turned to the *Inside Story* team:

"Excuse me a moment," he told them. "That's my cue."

Smoothly, he moved to a live mike.

"Dr. Helmut Borodine *will not return* until we have met tonight's pledge goal. Isn't it about time you thanked the professor for enlightening you? Pledge generously so that we may greet the Brothers from Space in the style they deserve: With pure gold!"

The in-studio monitors switched from a shot of the empty chair to the opening titles of the fifties sci-fi movie *The Man From Planet X*. Brother Ralph came back towards Peter and Tom, motioning them to follow him.

"This way," he said. "Let's see if Doc feels like talking."

* * *

They found the professor dressed in an expensive silk robe, a hand towel around his neck. He was smoking a cigar and, as usual, he looked irritable.

"Might as well tell you up front, I don't like your show," he declared.

Two pretty young women who were described as Borodine's research assistants were watching his every move with reverence. One of them had prepared his power drink, a mixture of orange juice with special vitamins. She presented it to him in a crystal glass.

"Didn't you appreciate our piece last week, about how the Vikings discovered California?" asked Tommy.

"Never seen it," he went on. "Don't have to. If it's not devoted to the Higher Truth, it's not worth watching."

Borodine got up and started to exit the room.

"What about our interview? We set it up on the phone . . ." Keller complained, running after him.

"Questions from the uninitiated are a damn waste of my time! You can film the next regression all you want, but there will be no interview."

Borodine walked off, silently followed by the two BGS girls. Thomas turned to Brother Ralph with amazement.

"Regression?" he asked. "What's he talking about now?"

"Regressive hypnosis. Completely free. A scientific service to those of our friends who have had Close Encounters of the Fourth and Fifth Type."

"You mean abductions?"

"We prefer to refer to these episodes as Enlightened Contacts. Let's follow the professor, or you'll be missing the induction."

"And who is the lucky subject?" inquired Keller, scratching notes on his pad. Ralph answered obligingly, looking up a schedule.

"A woman. It happened on Long Island, just a few days ago. Her name is Rachel Rand."

* * *

Moving quietly into the room with his video camera on his shoulder, Tom Archer started filming Borodine's hypnosis session with Rachel, who was already locked in a deep hypnotic state.

"My car is dead ... Engine ... won't start ..." she mumbled. Keller and Brother Ralph watched from across the room. They noticed Rachel's breathing speeding up; suddenly, she sounded very afraid.

"There are men ... men in the road!"

"Are they tall and blond, Rachel?" asked Borodine breathlessly.

"No ..." she intoned slowly, as if drawn far from the room by overwhelming terror.

"But they have blue eyes, don't they? Like the ones in that book I showed you?"

She mumbled a phrase they couldn't catch, her face convulsed, twitching with fear. Borodine leaned closer to her, impervious to her distress, eagerly questioning her:

"Which planet are they from? What's their message for mankind?"

Rachel's breathing continued to race; her features twisted. In her mind she was living the scene again, the misshapen heads pressed against the window of the car, the hairless faces, the emotionless black eyes that looked like pools of horror. There were three of them, and they came for her ... She screamed.

"Rachel! Hear me now. It's only a memory, you're safe here," Borodine assured her.

But she was so terrified she only heard him as a faraway voice. She blinked rapidly, her mind reeling back to that night on the road, when dwarfish beings dragged her against her will, down the highway toward a brightly lit object hovering just above the ground ...

She had entered a dream-like wall of light. She was lying on a table in a strange domed room. Cold alien faces were leaning over her. They grasped shiny instruments and made precise, razor-thin cuts along the inside of her forearms.

"No ... No!! They're cutting me ... not again ... they have no right!" She wept.

Ignoring her sobs, Borodine leaned forward, placing a pyramid on her head and running a compass over her thighs.

"What's he doing?" an amazed Peter asked Brother Ralph in a low voice.

"Checking for magnetic radiation," came the answer in a respectful whisper.

"Oh. Right," said Keller, repressing his revulsion, a spasm of utter disgust.

"Those are not true images of our Space Brothers, my child," Borodine intoned reproachingly. "My instruments are registering negative vibrations. Don't let your pain outshine the truth. I will work with you. We must break down the wall of fear that is distorting your contact."

He produced a crystal pendulum, swinging it over her chest.

"The stone doesn't lie. It tells me that the images you see are false. Our Visitors are good, they mean well. They come across the void to save our planet."

Rachel shuddered as her subconscious mind took in Borodine's suggestions. Keller couldn't help but smirk and shake his head at the pathetic spectacle he was witnessing.

"We've seen enough," he said, motioning to Tom Archer, who

stopped filming. He took Ralph aside. "Let's wrap this up. You said you had some background footage we could use?"

"The media center is just down the hall," the man answered obligingly. "You're welcome to pick whatever you want."

* * *

Brother Ralph, carrying an armload of videotapes, dumped them on a table near Thomas and Keller. They had begun reviewing the vast video archives of the Brotherhood, and they were using one of the half-dozen recorders in the room to watch Borodine give an impassioned outdoor speech to a large crowd of followers.

"Ah, I believe that's from last summer's Four Corners Festival," Ralph observed when he saw the picture. "One of the Professor's finest speeches, *Biblical Prophecy and the End Times*. Will you be able to attend this year's festival next week in New Mexico?"

"I doubt it," Peter said.

"Too bad. You'd have learned much from direct contact with our members. They come from all over the world, you know, just for this event. But these video recordings may help you."

He motioned to the rows of videotapes on the shelves.

"For your edification, these are more tapes from the Brotherhood's library, covering many aspects of UFO reality. You can copy whatever you like. We're working on a CD-ROM, but it won't be ready till Christmas."

* * *

Brother Ralph picked up a tape at random, popped it in and turned away. As he left, Rachel quietly entered the room, unnoticed by the two journalists. She wiped her eyes with her handkerchief, composed herself, and remained standing in the background.

On the screen they saw a man sitting in a chair, talking to the camera. His real voice was garbled by an electronic scrambler and his face was totally blurred by fuzzy moving squares. Below his face a line of type read "Mr. X, Former Top Government Scientist."

"It happened last summer in a remote part of Nevada," the man was saying in a tone of utter discouragement. "Top scientific specialists had been brought together to work on Star Wars technology. It was secret, a black program. But the project that paid our salary was only a cover. The truth is, they were trying to capture a complete, working saucer. They had already recovered bits of hardware and humanoid bodies, stored there in an underground base. The debris was recovered back in the forties . . . and . . . and I know it's real because I saw it with my own eyes!"

Nervously, the man went on: "And I saw . . . I saw other things going on out there . . . Things that would shock the world and change everything. Religion, politics. Things that shouldn't be!"

"Oh brother . . ." uttered Keller.

"I cannot reveal my identity. There are forces that would kill to keep the real facts from the public, and from the media," the mystery scientist went on, ominously, black squares dancing in front of his face.

"Not very convincing," Keller said. "I've had enough of that. Nothing tangible, no real data. How do we know this guy isn't just an actor, some starving extra they found on the street, looking for a bit part?"

He switched off the tape.

"Or maybe it's just another lunatic, this so-called Mr. X," he told Thomas. "Like this guy Borodine, and good Brother Ralph, and the girl we just filmed, and all the others. That girl, by the way, she doesn't need a hypnotist, she needs a good shrink."

Rachel took a quick step forward.

"You don't believe my pain is real? Or do you think I haven't suffered enough?"

Keller spun around, embarrassed to find her in the room. She moved towards him, her emotions seething under the surface.

"Next time, if you've got something to say, say it to my face, Mister. Because you don't know a damn thing about what happened to me!"

"I think . . . I know an active imagination when I see one," Peter said quietly.

"Did my imagination do this?" She challenged him.

Tears sprang from her eyes. She opened her sleeve, revealing long, straight, razor-thin scars on the inside of her elbow, down to her wrist.

* * *

Rachel was still furious as she walked swiftly across the parking lot.

"Miss Rand! Rachel!"

She turned to find Keller hurrying after her.

"I'm sorry," he said. "Look . . . you're right, OK? I made fun of something I knew nothing about."

Rachel kept walking. Keller touched her shoulder, stopping her in her tracks.

"Let me make it up to you. I want to do an interview . . ."

"So you can ridicule me coast-to-coast? No, thank you."

She started to walk away again, but he blocked her path.

"Why are you here? Why did you come to The Brotherhood? You've got to talk to me."

"These people support me," said Rachel. "Can't you understand that? They're the only ones who don't think I'm crazy." Somberly, stubbornly, she added: "I need to find out who those damned aliens are."

"Suppose we could help you?" said Peter. "You can think what you want about our show, but we reach forty million people every day. Don't you think they could support you, too, if they only knew your story?"

She hesitated. She turned towards him and stared into his eyes.

"All right," she finally said. "I've come this far. Maybe you can help me. Or you will curse me, some day, for dragging you into all this."

Keller didn't pause to consider her words. He motioned for Thomas to start filming.

CHAPTER SEVEN

It was four o'clock the next day, Wednesday afternoon, before Peter had his first chance to review what Rachel had told him. He was back at the editing machine, watching the interview she had finally granted him in the Brotherhood's parking lot. She came across as sensitive and intelligent.

"Being an abductee separates you from the rest of society," she was saying. "You can never feel safe, because you know it's never over. The images remain in your mind forever."

Ben entered the room, cigar smoke billowing into the air, and planted himself in front of the monitor.

"I've decided to dedicate my life to getting to the bottom of this mystery," the young woman was saying. "It's the only way I'll ever find peace."

"Just saw the roughcut you gave me," Ben interrupted, reaching over Peter's shoulder, hitting the Freeze button. "Most of it I love: the space lady who thinks she's the queen of some planet, the crazy paranoid scientist, all the Borodine shit, that's great stuff. And I love the hypnosis, especially the part where that crackpot swings the pendulum over her tits. But your interview is out."

He pointed to the screen:

"That's her, isn't it? Your piece is too long. Cut out the interview section."

He pivoted and went out, but Keller bolted out of his chair and followed him across the main office. Co-workers looked up and caught the heated exchange:

"That woman was crying real tears, Ben. Something definitely happened to her. It's the only meaningful thing in the whole god-damn piece."

"The hell you say! It's boring, it's irrelevant, it's out. I don't care what she's dedicating her life to. And I ain't gonna fucking argue about it!"

Keller turned, scowling as he brushed past startled co-workers. Thomas caught him as he reached the door leading to the elevators.

"Pete, we gotta finish the cut."

"Let Ben do it. I'm sick of being treated like a cub reporter."

"That's not his job. It's your job, your piece," Thomas reminded him.

"Yeah, in name only."

"Look, man, this isn't *Sixty Minutes*; we're not winning any Pulitzers here!"

"That's no reason to turn out bullshit."

"People want entertainment; that's what we do. They get enough bad news on all the other channels. The world's falling apart, in case you haven't noticed. There's big trouble in Mexico, Russia is on the verge of nuclear war with Ukraine. Don't you think people need something to take their minds off that stuff?"

"This isn't entertainment, it's plain trash."

"You look down on what we do, Pete. You don't even try. I've been carrying you, but Ben won't have that very long. Next time he'll give you a swift kick in the butt."

"Sorry Tommy, I gotta get outa here and get some fresh air. It stinks in here," Peter said as the elevator door opened up and swallowed him.

* * *

The subway train rattled to a stop in a hiss of brakes and a reverberating clang of obsolete machinery. *Grand Central Station* read the sign, nearly obliterated with hastily scrawled phrases and obscene drawings. A lone figure stood on the platform, a middle-aged man with a gray mustache who climbed wearily into the graffiti-covered carriage, seemed to hesitate and finally decided to land in the seat facing Keller, who paid no attention to him.

"Shouldn't have stayed in that bar so late," he was thinking. "What's going wrong with my life?"

Thoughts of Kathleen came drifting by. He chased them away. There had been other women in his life since the divorce, but few had been able to cope with his temper. Until Lisa came along, a very pleasant Puerto-Rican lady who didn't allow his wild mood swings to affect her healthy sense of humor. She must be waiting for him in his apartment right now. He raised a hand to his forehead, trying to chase away the dark thoughts. He should have been home an hour ago.

"The world is unpredictable," muttered the figure in front of him.

"Nine thirty at night," thought Keller, "and I'm drying out in this garbage can of a subway car. The only people who talk to me are either drunk or crazy."

The grayish, indistinct figure spoke again.

"I've got a friend, he says the whole planet is changing now, fast as lightning. Only one more year till the millennium."

"I wish this guy would shut up," thought Keller, his brain throbbing. "Next thing you know, he'll be quoting Nostradamus."

Somewhere in the dark tunnel there were screeches. The train lurched forward, then slowed to a crawl.

"My friend is a Kremlinologist, a former Soviet expert. He has trouble finding another line of work."

"Don't we all?" Keller said, hoping to put an end to the stranger's monologue.

He was apathetic. He had too many problems of his own to worry about some weirdo in the subway who fantasized about the Soviets. Communism had vanished into the trashcan of history. Good riddance.

"The dawn of a New Age," the man went on, his voice rising. "Old enemies are new friends. There is plenty of room for new mythologies. Who would have guessed?"

Keller tried to focus on the lonely silhouette. The man was in his early sixties. Clean. A black hat and an overcoat. A pale, wrinkled face where the gray mustache put a touch of innocence and wisdom. He was not looking at Keller at all. He was reading the *Daily News.*

"After forty years of hating our guts, you think the Ruskies simply changed their minds?" He went on, as if he were speaking to himself, or reading the newspaper aloud. "Do you really think ol' Gorby just woke up one morning, took a look at the graceful curves of Raisa sleeping next to him in her nightgown, and said *Let freedom ring?*"

"Who knows?" Keller finally said. "People change."

"Ah, but dictators don't. Not so fast. Not so completely."

The man had taken his stare off the paper and now he looked directly at Keller with his piercing gray eyes. The fellow might be crazy, thought Peter, but at least he wasn't hopelessly drunk.

The car shuddered again as it moved along a curve. It leaned at a twenty-degree angle while every piece of metal in the train seemed to complain with loud screeches.

"Well, it happened," Peter said to stop the conversation once and for all. But the man with the gray mustache did not give up.

"Question is, why?" he said.

"I suppose you're gonna tell me," Keller replied wearily.

The old man folded his paper and leaned forward, looking Peter in the eye. The train had righted itself, but the lights were blinking ominously, on and off, like strobes in a dreary Manhattan loft party.

"Remember good old Ronnie Reagan? Remember when he met Gorbachev in Iceland?" he said with distinct words, spoken loud enough to be heard over the roar of the careening train. "Back when they were still the Evil Empire? Or have you already forgotten, like all the other monkeys down here?"

He gave a wicked smile. Keller remained silent.

"Well, what if Ronnie had a big stick? A secret weapon so powerful the Soviets had no choice but to submit, across the board?"

The smile faded. Keller had a chilling sensation that the old man spoke not from fantasy but from personal experience. There was no stopping him now.

"Reagan said, 'We'll allow you to lose gracefully. You will dismantle communism and the Warsaw Pact. You will tear down the Berlin Wall. You will apologize for your meddling. You will establish free elections, and a free press inside the Soviet Union.' Just like that."

He laughed, apparently elated by some private recollection.

"Well ol' Gorby nearly choked. 'Cheer up,' says Mister President. 'You're gonna be the Man of the Decade!' But they don't just agree like that. 'We'll need proof,' the Soviets say. 'Of course you will,' replies Ronnie-boy."

The train screeched to a painful stop at another seedy station. This time Keller was actually relieved to see that no one came on board. The doors closed in a slam of metal, and the car lumbered down the tunnel again, along tracks that sloped downward as if on their way to Hell.

Keller didn't want to admit it, but now he was thoroughly fascinated by the man in front of him, and by the bizarre story he told.

"So one fine day we let the Ruskies visit a certain place in Nevada. And a few hours later a bunch of hotels in Vegas swayed at high noon. The local papers called it an unannounced nuclear test."

The train was slowing down, ready to enter another station. The man got up, folded his newspaper, winked at Peter Keller.

"Well, the Reds turned white. Damn, you should've seen their faces when they left. Walking dead men. Because you see, it wasn't a nuclear test. It was a thousand times better. We scared the shit out of them. They had their proof, and the rest is history."

"Quite a tale," said Peter. "You make it sound like it really happened."

"It did," said the man with a grin.

He handed Keller a business card.

"*Nice talking to you, Pete,*" he said as he left the train.

The doors closed and the subway car rumbled on. Peter looked down at the business card, which read "A. Smith." All he could do was to shake his head and chuckle. He leaned close to the window to catch a last glimpse of the terminal. It was deserted. There was no sign of the man in the black coat who had just stepped off.

CHAPTER EIGHT

Nowhere in New York City can you get away from the ominous rumble of the metropolis. It lurks in the ripples of the water left in the glass, in the slight stir of the drapes, in every tiny tremor of the window panes when you touch them. Keller had once interviewed a scientist, an electronic expert who had rigged up a complicated device he called a "multichannel spectral analyzer." He had deployed the antenna in his living room, and he had shown Keller lime-green curves on the screen, representing all the low-frequency waves he said were booming through every dead and living thing over a fifty-mile radius.

Keller was aware of the rumble of the city that never sleeps because he was sitting on the floor, a glass of beer in his hand, and the wall against which he was resting his back was vibrating.

He heard Lisa turning on the shower in the bathroom.

The phone rang. He rose to answer it. His daughter.

"Hello? Oh hi, Squid." He was happy to hear Sara's voice. "What's shaking, Sweetie?"

He could see her in his mind, calling from the house in Westchester County, *his house,* with Kathleen and Gérard on the couch in the background, probably eating tofu.

"What are you doing still up at this hour?"

"We're watching your segment, Daddy, 'bout the space people."

"Oh yeah, it's running tonight, isn't it? I'd forgotten that."

He hit a button on the clicker, and his own living room set came to life. The booming voice of Thomas Archer commented on a lecture by Borodine, who gesticulated in front of a rapt audience.

"Borodine claims his life was changed by a roadside close

encounter with blond, blue-eyed spacemen who specifically chose him to spread the word about the degradation of the environment, the need for world peace and the coming ecological catastrophe. At the time, Doctor Borodine was a humble worker in Hoboken."

The camera showed an old photo of a much younger Borodine holding an acetylene torch. His welder's mask, which was pulled halfway up, made him look like an outer-space invader himself, but underneath it revealed a baby-faced man with chubby jowls.

"Great show, Dad," Sara went on. "It's very spiritual. I want to watch it to the end. I'll call you tomorrow and we'll talk it over."

"Right honey. I love you."

"Love you too, Daddy."

Keller hung up the phone. He saw Lisa coming out of the bathroom, a purple towel wrapped around her. She wasn't looking at him. Her eyes were glued to the television set. She wore the towel like a Mediterranean empress might have worn a toga, the resemblance strengthened by her Hispanic profile and the heavy makeup she used on her eyes. Her long black hair tumbled over her bare shoulders in graceful, heavy curls. She adjusted her glasses on her nose with one hand and held on to the towel with the other as she watched Brother Ralph grinning from yet another old, yellowish photograph. There were rolls of carpet behind him in the background.

"Borodine's sidekick is Ralph Augustine, a successful businessman once known as the Carpet King . . ." Thomas's voice went on.

"Where do you find these people?" asked Lisa with admiration.

She spoke with an easy grace, as if she found herself in an elegant salon instead of standing almost naked in a bachelor pad in Manhattan, her feet leaving small patches of water on the green carpet as she walked.

The jovial Brother Ralph spoke directly to the camera.

"On several occasions I've driven with the Doctor to the desert. And I've seen the lights come down, right where he said they would."

"How do you suppose he knew?" said Thomas's voice from off-camera.

"The Voices," Brother Ralph answered without the slightest hesitation. "He hears them directly inside his head. Doctor Borodine is a modern-day prophet, sent to warn us of the approaching Eschaton, as the Bible says. The Voices speak to him, like they did to Moses."

"Yeah, sure they do," said Keller with disgust, pressing the clicker, turning off the television set.

Lisa protested: "Hey, that was really interesting. What's an Eschaton?"

"Not you, too!" he said, pulling her body towards him, tugging playfully at the towel, kissing her. "C'mon, let's go to bed."

* * *

The bedside clock read 3:45 A.M. Peter and Lisa were asleep. She was on her belly, her left arm thrown around him with abandon, her black hair spilling over the covers.

The phone rang. She stirred. Peter fumbled for the phone and the light.

"Oh, my head," she complained, turning around, moving away from him with a slow, swimming motion. "Do you have any aspirin?"

He did not answer. He grabbed the phone like someone jolted from a nightmare, someone who thinks he is still dreaming.

Lisa groggily headed for the bathroom while he picked up the receiver.

"I'm extremely disappointed in you, Mr. Keller," a male voice was saying at the end of the line. "There was a good story there and you missed it, just like the rest of them."

"Who is this?" asked Peter, tired and annoyed. "Chad? Tommy? Is that you?"

The caller continued as if he wasn't aware of the question.

"Those voices he hears in his head? *They're our voices, our words.*"

"That's just fascinating. Really. Can I go back to sleep now?"

"It happened right after the war. My group was formed to find out what was invading our skies. At first we thought they were Russians, or maybe a splinter group of Nazi scientists in some weird corner of the Amazon. We were afraid of another Pearl Harbor, because of the climate in which we were born."

Keller was fully awake now. He was listening intently.

"After all these years, fear is still the spur," the voice went on.

"Surely you realize it's almost four o'clock in the fucking morning."

"Of course they weren't Russian. And fifty years later the secret is safe."

"What secret?" asked Peter, propping himself up on the pillow, holding the phone with both hands.

"A plane crash will be announced tomorrow in Nevada, Mr. Keller. A pilot has died because of what he's seen. He might have told his wife, or his girlfriend, or his son some day, and they can't allow that. Check the wire in the afternoon. This man had to die because he knew UFOs are real. You understand me, Mr. Keller? *Real.* Oh . . . and tell the girl the aspirin is in the bottom drawer."

The line went dead. Dazed and confused, Keller got up and went to the bathroom. He found Lisa rummaging through the medicine cabinet.

"Where's the aspirin?" she asked.

"Why did you say that?"

"Because I've got a splitting headache, why do you think?"

"It's on the top shelf. I always put it there."

She shook her head: "You find it."

There was no aspirin in the cabinet. Reluctantly, he yanked open the bottom drawer, and found it there. Lisa was startled as he bolted to the bathroom window and looked out into the night. His eyes eagerly scanned the rooftops, the windows of nearby buildings, and the streets of Manhattan below. All seemed normal, yet Peter Keller could not shake the feeling that he was being watched, that something ominous and hateful had entered his life.

* * *

"Don't look at me," said Holly with a shrug. "I was with my boyfriend last night."

"How about you, Chad?" asked Keller with an angry, accusing finger. The Australian laughed in his face.

"Believe me, I've got better things to do than play jokes on the likes of you."

"Oh yeah? Like what?" replied Peter.

"Like lots of things: spitting in the East River; or washing my socks." Chad Burke smiled ironically, showing an impeccable row of big, bright teeth. "Washing my socks is a lot more important than playing jokes on Mr. Peter Keller any day, believe me."

Holly giggled: "Ah, but were you responsible for hiding the aspirin?"

The whole group laughed.

A young black man entered the room, interrupting Keller's interrogations.

"Package for you, Pete. I put it on your desk."

Reluctantly Peter walked off, but not without a last word to Chad: "You've got guilty eyes. I'd watch it if I were you."

"This is infantile. You're the last bloke I'd call, day or night." Cheerily turning to the others he added, in a voice loud enough to be heard by everybody in the room, "Why waste time on a lost cause?"

Keller reached his desk, looked at the manila envelope which simply bore the words "P. Keller," and turned to the assistant:

"Where'd this come from?"

"Dunno, somebody left it at reception for you."

Before he could open it, one of the secretaries called to him from a teletype machine.

"Hey Pete, take a look at this."

Putting the unopened manila envelope under his arm, he joined her next to the machine, which was whirring and clanking away.

"You said something about a plane crash, right? Well, it's nothing big, but ..."

Keller read aloud from the long paper.

"An F-16 jet fighter exploded in a ball of fire over the rugged Groom Lake Mountains of southeastern Nevada on Tuesday. The pilot, Captain James Hall, was killed instantly, according to Air Force sources. The cause of the crash is under investigation."

"Is that what you were waiting for, Pete?" she asked. "Pete?"

But Keller did not answer. He went back to his desk and tore open the package.

It contained a single sheet of white paper, with a typed message: *Thursday midnight. Hudson Park. Seeing is believing.*

"What the hell is going on?" Keller muttered to himself.

* * *

The Jersey side of the Hudson River used to be a desolate place of lethal swamps and poisoned ponds, where the birds never sang. Even insects died in the ooze from dozens of chemical plants and refineries that fed New York City's hunger and thirst for toxic products. Yet someone must have decided to bring a little bit of quality of life to the neighborhood, because a park had been developed there, at a great expense in clean soil, young trees and tons of fertilizer. Even a few birds had come back.

Keller was there at the appointed hour, walking around in circles, his hands thrust down in his jacket pockets. He saw nothing unusual. The park was dark and quiet. He passed two homeless people sleeping in the bushes.

It was getting cold. Peter looked at his watch. Seventeen minutes past midnight—nothing. Of course it was another blind alley. He felt very foolish for coming over in the first place. He sighed, and decided it was time to leave. Another failure.

Keller was walking across the neatly cut grass when he sensed someone coming at him from the side. He froze, startled, as a powerfully built man in a hooded sweatsuit ran by, jogging across the park at a brisk pace, nearly hitting him.

"Jesus, pal, watch where the hell you're going!"

The jogging man kept going, but turned his hooded head to look at Keller. And at that precise moment the entire area exploded with sparks, as though a huge burst of electricity had hit the ground just behind Peter. Visibly staggered, the jogger was surrounded in eerie blue light. His face showed terror and awe. Keller spun around, instinctively afraid of what he would see.

A brilliant object was rising above the treeline.

It was a wingless device, a huge metallic disk, clearly under intelligent control. It emitted intense blue and green pulsing lights, and it floated above Peter in total silence.

Keller shielded his eyes with the back of his arm and cringed. It looked as if the enormous craft were about to land. He ran for the shelter of a nearby tree, feeling the hair stand up on the back of his neck.

The awesome device came down to within a few feet of the ground, rocking gently like a falling leaf, hesitating, as if looking for a convenient spot. Then, suddenly, it streaked straight up, moving faster than the eye could follow. It stopped again, frozen, perhaps half a mile away. For a moment it hovered high over the park, in plain view of the surrounding area, and eventually zoomed away, crossing the horizon at breakneck speed. It was out of sight in seconds.

Keller realized he was holding onto the tree and sweating profusely. He breathed hard, feeling dizzy. He looked across the park, but the jogger was nowhere to be seen.

Keller spotted one of the derelicts who were sleeping on a nearby bench. From the looks of it, he was just as shell-shocked as Peter.

"Did you see it?" Peter asked breathlessly. "Did you?"

The man was dazed and frightened, unable to speak. Keller grabbed him by his tattered jacket and shook him, a smile of realization spreading across his face.

"They knew ... Sweet Jesus ... They knew it was going to be here! Don't you get it? *Those fuckers knew exactly what was going to happen.*"

PART TWO

THE ROGUES

In warfare, truth is so precious that she should always be attended by a bodyguard of lies.

Winston Churchill

CHAPTER NINE

On Friday morning the production meeting was already in progress when Keller rushed into the offices of Inside Story, pushed open the large mahogany door with a decisive motion and stepped into the room. He found Ben Simon sitting at the end of a long table, a stunning view of Manhattan filling the picture window behind him. All producers and reporters were present. The single conspicuously empty chair belonged to Keller.

"Nice of you to join us, Pete. You still work here, don't you?" asked Ben.

Keller ignored the irony. He spread his arms wide.

"Something really incredible has happened. I'm onto something big. You won't believe how huge this is," he said breathlessly.

"Gee, that's swell, Peter, but I've just given Tommy your next story."

"You don't understand . . ."

"No, you're the one who doesn't understand!" yelled Ben Simon, cutting him off. "You're doing 'celebrity secrets' next week, my friend. So if your scoop's about the First Lady being a wrestler in drag, fine. If not, just sit down and shut up."

"Ben, for Chrissakes, this could be the story of the century. And get this," he added with an excited laugh. "It's about UFOs, can you believe it? Real UFOs!"

Ben looked around at the others, as if wondering why he had to deal with such a crazy man so early in the morning. He lowered his head into his hands as an excited Keller began pacing before his co-workers.

"I think it started on the subway ..." he said, rubbing his face.

He went on for ten, fifteen minutes, recounting every detail. He concluded: "The thing was absolutely unearthly. But that's not the most important part ..."

Keller looked around the room. Chad was plainly amused, Thomas was uneasy, and the others were fidgeting.

"Don't you get it? That phone call actually predicted the sighting!"

"So who was it that called?" Ben pressed him. "Deep Throat? An alien? A giant tarantula from Pluto? I thought you had an unlisted number ..."

Chad had placed two spoons over his eyes.

"Calling Keller, calling Keller!" He intoned in a nasal, robot-like voice, to a chorus of uneasy laughter.

"Pete, Pete, Pete, whaddya got to back this up? Something tangible," Ben insisted.

Keller knew he had no proof. He felt angry at the whole lot of them, and even angrier at himself for trusting them to take him seriously.

"Look, get this guy on the phone; find him, for God's sake, let me check him out, then I'll make a judgment." Ben's manner was conciliatory.

"I told you, I don't know who they are," Keller confessed dejectedly.

"It's just as well, Ben. I think Mars is a toll call," Chad said.

"What about the park last night?" asked Thomas seriously, amidst the laughter. "If Pete says he saw something ..."

"Sure he saw something," Chad broke in, cheerful as usual when he was saying something nasty, "and he's not the only one, either. It's on the wire this morning."

Holly jumped: "You mean other people saw the saucer?"

The man shrugged.

"They thought they did, twenty minutes past midnight last night, all along the East River. But it was a meteor shower."

"Oh man, that's bullshit," Keller protested, throwing his hands up in the air.

"He's right Pete, I heard about it too," said Simon's secretary. "It was all over the radio news."

"Give me a break!" he yelled. "There's no way that's what I saw! It practically landed, less than a hundred feet away!"

"Face it, the sighting's been explained. Some astronomer from Cornell University was even quoted about it . . ." Chad insisted.

"Fuck Cornell! And fuck you, too!"

Keller angrily stormed out of the room. Thomas made a motion to go after his friend, but Holly gently took him by the arm.

"Let him cool down, Tommy. Let him be."

* * *

Keller was dejectedly downing his second drink at the corner saloon when a tall Oriental woman seemed to materialize out of the shadows. She wore a dark business suit, a blue necklace and matching earrings. She carried a blue folder, which she deposited on the bar as she sat down next to him.

"Kinda early on a Friday to be drinking hard liquor, isn't it?" she asked.

A little surprised, he looked her over as he signaled for another.

"My name's Junko," she said pleasantly, offering her hand.

"What'd you come here for . . . the peanuts?"

"Actually . . . I came over to discuss astronomy. Meteors, in particular."

The next drink had been served, but Keller hadn't noticed. His eyes were now riveted on the willowy stranger. He couldn't decide her age. Forty? Forty-five?

"You were there? You saw it?" He whispered in amazement.

She laughed.

"I didn't have to see it. I knew precisely what was going to happen."

"What else do you know?"

Suddenly she dropped the chatter and became all business.

"Well . . . for starters, I know a lot about you," she began with a hint of a smile. She picked up the folder, opened it and started reading from a typed sheet.

"It says here: Columbia. Journalism, class of '81. *Rolling Stone, Harper's*—seems you were pretty good in those days, and you knew it, didn't you? But you got a little careless, Pete. A little too sure of yourself. One bad move, a pity. A career destroyed over one lousy, unverified piece."

Keller turned away, disquieted by this conversation. He downed his drink, images flooding back to his mind, thoughts he had wanted to push down forever into the darkest corner of his memory.

The stranger was right.

There had been a time when Peter Keller was a respected political journalist. Then he followed up a tip from a confidential source who knew the husband of a prominent California politician. The candidate was a woman who seemed destined for a brilliant career, but Peter was informed of serious irregularities in the couple's financial affairs, a multi-million dollar resort development in Costa Rica that appeared to have Mafia connections.

For weeks he had traced every piece of data, checked every fact—or so he thought. He interviewed a convicted killer at Folsom Prison, a former member of a drug ring who told him how he had put the real-estate deal together with laundered money.

The exposé he published destroyed the candidate. She tried to commit suicide by jumping off the Golden Gate Bridge. Her life was saved by a passing motorist who pulled her from the railing, but her spirits were broken and she never recovered enough confidence to run for office again. Yet her supporters kept on fighting to rehabilitate her.

When all the facts came out, the flaws in Peter's research became apparent. The convict had lied. Other people had been bought out. The documents were forgeries. Someone had made up the story to ingratiate himself with friends of the governor. Keller was fired, and had to publish an apology for his article. He never found out who had fed him the tainted information in the first place.

Now a perfect stranger had walked up to him in a nondescript bar to dredge up that story. And she wasn't through with him.

"Why do you figure your wife left you? Because you fucked up, or because you couldn't admit you fucked up? Or was it the drinking? The running around?"

An incredulous Keller looked at her with murderous eyes.

"I know, Pete, if I were a man you'd punch me. Well, I've been punched plenty of times, go ahead if it'll make you feel better."

Keller looked away, composed himself.

"I take it you're friends with that ghoul from the train, the one who called me at four in the morning."

"What makes you say that?"

"You both speak fluent double-talk, and you know a little about me, each of you. Like where I got my diploma, who I'm sleeping with . . . real impressive stuff. Strategic information. You ought to call the Pentagon right away."

She just kept staring at him, waiting for his anger to blow over.

He frowned: "It seems you people are trying to tell me something. A dead pilot, the Soviet debacle. Now a close encounter. Well, so what? What's it got to do with me? What's the connection?"

"Don't you already know? What kind of a journalist are you? You've seen it with your own eyes, haven't you?"

As Keller was pondering her question, she abruptly got up. Before he could blink, she was off, heading for the exit.

"Wait!" he called.

But she kept on going. In an instant she was out the door. Keller spun around on his stool, digging in his pockets for cash. He laid a twenty on the counter, and hurried after her without waiting for his change.

* * *

The parking lot was deserted when Keller ran out. He scanned the area to the left and to the right, but the woman was nowhere to be found. He began to walk across the lot toward the street.

A black sedan with dark, tinted windows turned the corner and pulled up in front of him; the front door was thrown open as a clear invitation to get in.

A man in a charcoal suit was behind the wheel, with Junko in the back seat. They looked at Keller with unreadable faces, patiently waiting for him to step inside. After a moment's hesitation, he did so. "My mother told me never to get into a car with strange men," he said. "Or strange women." He tried to control his voice, to sound sure of himself.

Keller shut the door. The driver was already hitting the accelerator.

"Where we going?"

"Wonderland. You're gonna like it."

The sedan wove in and out of traffic, expertly cruising across Manhattan.

Keller leaned forward and brazenly opened the glove compartment. He pulled out the car registration and studied the name on it: "Colonel Richard Trent," it said. But there was no listed address.

"Always the intrepid reporter, eh, Mr. Keller? I wouldn't pay much attention to that name if I were you; it's different in every state."

"And that makes you some kind of spook, I suppose."

"Such a nasty word! Let's just say that I deal in information. Just as you do, Mr. Keller."

"Information . . ." Peter repeated with a sarcastic edge to his voice.

"Yes, Mr. Keller. A most valuable commodity. There's no such thing as privacy anymore: We can enter your bank records, trace your credit cards, tap your phone, bug your house, and you'd never know the difference. It's easily done; it's done all the time."

"So you've invaded my privacy. Big deal. Real question is, why? What do you want from me?"

Trent looked at the woman over his shoulder.

"Haven't you told him, Junko?"

"He doesn't seem to grasp the whole situation quite yet," she answered. And she went on, detaching every word: "We want to offer you something priceless, Mr. Keller. A chance to regain your self-respect."

"What makes you think I've lost it?"

"Come on, Pete, you work for *Inside Story,* remember? We can help make you a real journalist again."

After a pause, Trent added: "What we have here, to quote your own words to your colleagues as recently as a few hours ago, is *the story of the century.*"

While Keller reflected on that, the car came to a stop at a busy Broadway intersection. Through the windshield, he could see masses of people milling on the sidewalks or crossing the street in front of the sedan.

"Take a look at the flock, Mr. Keller: Conformity is in, and individuality is very scary. We live in the twisted age of opinion polls and Nielsen ratings, of late-night Internet chats with harmless strangers and interactive computer sex. Why leave your house on election day when the local news, the night before, has already told you who the winner will be? No wonder people don't vote, no wonder they don't care."

"Are you a sociologist as well, Mr. Trent?" asked Keller drily.

Trent smiled at the question. Red turned to green. He drove on.

"More a philosopher, I suppose."

"And I take it your philosophy doesn't include much love for democracy."

"There you're very much mistaken, Mr. Keller. I would give anything to live in a democracy. But what we have here ceased to be a democracy decades ago. The founding fathers based our system of government on the free vote of an informed citizenry, remember? Well, the people you see in front of you haven't been informed on any of the really important issues for a long, long time."

"Don't be too sure. Humans are smart animals . . ."

Trent cut him short: "Humans are sheep. They'll follow you anywhere," he said drily. "The trick is to be the shepherd. Always the shepherd."

* * *

The car kept driving down a long tunnel, then along what Peter recognized as the Gowanus Expressway on Long Island. Soon he saw the Varrazano Bridge towering over him. They reached the Staten Island side, and rolled to a stop in an area just beyond Fort Wadsworth. The place was run down and garbage-strewn, the sidewalks reeked of decay, and the old constructions around them were stained with graffiti. Pete glimpsed a patch of dirty river between the buildings. Ships and tugboats howled in the distance.

Trent and the woman led an intrigued Keller along the street toward the side of a massive, weathered brownish structure. He saw it had once been an outdoor stadium used for football, but now abandoned. Trent pulled out a wire cutter and expertly sliced through some rusted chains. Then he and Junko led the way up a rotting concrete staircase into the stadium.

Flanked by his two guides, Pete walked out onto the weed-infested field; he stopped in his tracks, scanning the dark stadium bleachers.

"So what gives? We gonna play football? Or just mow the stupid lawn?"

The cockiness was to reassure himself. But when he looked around, Junko and Trent were no longer beside him. Keller noticed that he was bathed in an eerie blue light. He spun, covering his eyes, the whole world swimming in electric blue.

The alien machine he had seen at Hudson Park was floating toward him again. It came on silently, flashing blue and green, growing larger and more ominous.

Keller felt intense heat when the wind blast from the saucer whipped up discarded scraps of paper all around him. He backed up, stumbling to the ground, instinctively shielding his face. The thing kept coming on.

Soon it was directly above him.

The light was hurting his eyes, making them water, but he could make out the rotating disk shape, and what seemed to be a polished metallic surface.

"Okay! Cut the beam," a male voice said on the loudspeakers.

The saucer vanished in an instant. Keller rubbed his face, trying to recover from the piercing blue light that had emanated from the disk. He looked around, blinking his eyes, and saw several figures approaching in silhouette.

The next voice he heard was that of Junko.

"Your meteor shower, Mr. Keller," she was saying. "As real as a dream, as solid as a smoke ring."

A small group of men and women were becoming discernible in the near-darkness. Someone turned on the handful of ancient spotlights that still towered high above the stadium. He saw Junko, Colonel Trent, a young bald fellow with glasses who wore a red T-shirt, the ghostly man in black from the subway, and a few others.

"Always a pleasure," said the man in black, shooting a knowing smile at Pete.

Keller was still staggered by the complex illusion he'd just witnessed. Confused, he looked again at the now-empty sky.

"But . . . how?" he stammered.

Junko pointed to a small but intricate machine that straddled the first row of bleachers.

"A simple holographic projector," said Colonel Trent. "The technology is six years old. We used it last week in Kansas, scared the piss out of some poor farmer. Five A.M., and something from space is landing in the cornfield. Made a few local papers. Even left a double ring imprinted in the landscape."

"No one believed him, of course."

"But eventually they will. That's the whole point."

"Sooner or later."

"But the heat . . . and the wind?"

"Parlor tricks, Hollywood magic," Trent answered with a shrug.

"So . . . none of that is real," Keller managed to mumble.

"What difference does it make, if people believe it? Does a unicorn have a horn in the middle of its forehead, Peter? Does it?"

"Well, . . . Yeah, what are you driving at? Of course it does."

"And is there such a thing as a unicorn?"

"Of course not."

"How can something be true and false at the same time, Peter? Don't you see? Ordinary logic doesn't apply to what we do. If you want to work with us you've got to leave it behind."

* * *

Keller took a moment to consider his experience and to calm himself. The whole group now stood before him. He scanned the rest of the stadium. He could see a handful of men, plainclothes military types, roaming the rafters with automatic weapons by their sides. They were watching the area, keeping a vigil. Keller sensed an underlying tension in their movements.

"Let's get on with it," the bald fellow said. "I don't plan to stay a minute longer than necessary."

Junko moved toward Keller:

"Lest you think we're nervous types, we have friends—"

"—enemies," the bald man interrupted. She ignored the remark.

"I said friends—who wouldn't approve of this meeting."

"Are you government?"

"Yes and no. The taxpayer foots part of the bill, if that's what you're asking. What difference does that make? We are part of an organization, each one of us a specialist in his or her own field. We are very, very good at staging simulations, like the one you saw, but our primary task is to collect information, the real thing."

"And the name of this group?"

She answered: "It has many names . . . and no name. We call it Alintel."

"Alintel.. Alien Intelligence?" Keller asked, slowly overcoming his initial feeling of awe.

"Give the boy a cookie," said the man in black. "You sure know how to pick 'em," he added, turning to Trent.

"*Alien Intelligence*... no shit. Who would form a unit like that?" Peter asked Junko.

"A president, many years ago. He designed Alintel as a black project: No records, no traces, no bureaucratic trails. There were only twenty-eight members at the beginning."

"And they studied UFOs?"

"They tried. God, how they tried! It was supposed to be a five-year study, ten at the most."

She broke into a wry smile:

"Well, here we are. It's still going, half a century later. The UFO phenomenon has turned out to be far more complex than anyone had dreamed."

"So you do believe it's real."

"Oh, we knew that from the start. It's not a question of belief. We had radar trackings, eyewitness accounts, photographic evidence. A few pieces of tantalizing hardware."

"That's nothing the public hasn't heard about—but no one who's anyone believes in those things anyway," Keller pointed out angrily.

Hell, even his own colleagues had told him as much. They had practically thrown him out of a production meeting.

"That shows what a good job we've done, don't you see? Placing a curtain of laughter and ridicule around the entire subject; making sure the real cases were buried while the crazy ones made the headlines in the tabloids."

"And the talk shows. Don't forget the talk shows," the man in black added.

"Alintel has rapid deployment teams, blue berets, set up throughout the country," Colonel Trent said. "When a real encounter takes place, they're assembled and ready to roll in minutes."

"You don't need to tell him everything," Junko protested.

"We agreed: No more secrets," insisted Trent. He turned back to Keller:

"Alintel knew that these craft were invading our airspace and taking people away, but we didn't have a clue as to where they were from, or why they were here."

A silence. Peter was dumbfounded.

"And then ... we found some debris," Trent concluded. "All broken up. But some of it still worked."

The man's face lit up as he recalled the long-ago discovery in the Southwest.

"It came crashing down in the desert. A rancher stumbled upon it."

"1947, right? In New Mexico? I read about it."

"No, Pete, what you read was just the cover story, a blind alley. The real pieces were picked up the year before."

"Jesus ... Then you people have had all the answers for over fifty years!"

"Wrong again, Pete. It was one piece of the puzzle, but not the smoking gun. You see ... we could never reassemble the thing. After years of trying in our labs, the hardware was taken from us and shipped somewhere else. We think it is in Nevada."

"With what was left of the crew," Junko added.

"You mean ..." Keller felt the blood drain from his face.

"Yes," she said. "The craft was broken up, but several of the Aliens were alive."

Before Peter could think of a question to ask, Trent noticed a commotion in the rafters. The bald fellow stepped between Keller and the man in black.

"We've got activity out front," he said.

The Alintel people were in motion within seconds, dashing off in different directions. Trent and the woman shepherded a confused

Keller toward an exit.

They followed a winding, rusting staircase to the street level. Keller noted their concerned expressions; heads down, they led the way across the decaying street and down a small slope toward the water's edge.

CHAPTER TEN

Trent and the woman finally stopped in a hollow by the banks of the dirty harbor. Discarded wood, newspapers and beer cans littered the ground; the Statue of Liberty loomed beyond the water in the distance.

"See anything?" the woman asked.

Trent looked back over his shoulder, a little breathless.

"I think we're all right," he replied, beads of sweat showing on his forehead.

"Scared, Trent?" Keller asked sarcastically, his anger rising again. "What is it with you people? One minute you're so damned arrogant, and the next you're afraid of your own shadows."

"Alintel has a long shadow indeed, Mr. Keller. You're the first outsider allowed to see the wizard behind the curtain. We've got to be careful; there's a lot at stake here."

"So you don't represent Alintel after all?"

"We're a rogue element. You might say we're in rebellion," Junko said.

"Why?"

"Because this time we're afraid. Something's happened, something important."

She came closer to him, her words ringing with intensity.

"In the past few days we've noticed changes in Alintel. Rumors began to fly about new information, a plasma gun they hoped could disable a Fastwalker in flight. There's even a possibility Alintel has captured a craft—a complete saucer this time, but we aren't allowed to see it. The data is kept away from us. Newcomers are being brought in—military men, scientific types. It's a shift in long-standing policy, with no apparent justification. We find ourselves being pushed out of the inner circle. Suddenly we don't know who we're working for any more."

"And the world has begun to change," added Colonel Trent.

* * *

The phone rang at the CIA, in the office of James O'Grady, Ph.D., a thirty-five-year-old scientist who had been assigned the task of looking after all the bizarre, undocumented and unwanted data that the Agency's vast collection networks were dredging up all the time. Such data didn't fit into any useful political or strategic category, yet it always merited some degree of attention.

"You never know when some innocuous ghost story or some silly occult belief may turn out to be significant," his boss had told him when he was assigned the job of "Keeper of the Weird."

"Look at Hitler's obsession with astrology; or the apparitions of the Holy Virgin in Croatia, a few years before Yugoslavia blew up in our face."

On this particular day O'Grady, a tall man with wire-rimmed glasses and a fine mustache, had already scanned a thick folder of National Security Agency telephone intercepts concerning French parapsychological experiments, and several diplomatic wires about Chinese expeditions into Tibet in search of the elusive Yeti, when the phone rang and he found himself talking to Bill Renslow, the White House Chief of Staff, whom he knew to be a close friend of President Henry Templeton.

"What can I do for you?" asked the young analyst, immediately on his guard. The Directorate of Science and Technology was not used to taking phone calls from 1600 Pennsylvania Avenue.

"I am told that you're the man in charge of analyzing strange occurrences," said the voice at the end of the line. "Does that include UFOs?"

"Not officially, sir."

"We're not speaking on the record here, young man," the voice said drily. "And you're not talking to CBS News. This is the White House, and I'm asking you, yes or no, if you keep up with UFO data."

"Yes, sir. I mean ... In public we deny it ... but I try to make sense of whatever crosses my desk. And most of it turns out to be garbage, naturally."

"Excellent. I'm sure you fellows do a fine job. I have a favor to ask you. Some friends of the president are putting a lot of pressure on us to look into all this UFO business. We have other fish to fry at the moment, of course. I'm sure you know as much as I do about what's going on in Mexico, where the social turmoil is ready to boil over. They might have a coup any day down there. However, these are friends of the president, influential people, Washington socialites, political contributors . . . I'm sure you understand."

"Yes, sir, of course."

It was hard to turn down polite requests from people who influenced vast sectors of public opinion, even when they freaked out over such absurd rumors as flying-saucer stories.

"We are setting up a briefing for Tuesday afternoon at 1500 hours. I'd like you to prepare a background paper. If you show your credentials to the Secret Service at the side entrance, they'll take you straight to the conference room."

* * *

Dr. James O'Grady hung up the phone and scanned the metal cabinets equipped with combination locks that lined one entire wall of his office. Trust the White House to call him up with such a bizarre request at five minutes to five on a Friday, when he was ready to leave, he thought. He would have to cancel his plans to visit his fiancée in Connecticut. And he would have to find out what information the Agency had on a subject that had generated far more controversy than hard data over the years. Judging by the masses of reports his elders had left behind, it appeared they had wavered about the nature of the objects. Now he had three days to draft a briefing for the chief executive. No time to start doing serious research, of course.

O'Grady was too good a scientist to believe in flying saucers, but he was too smart an intelligence analyst to present himself unprepared before the president of the United States. So he called up the information science department, and initiated a database search of all recent foreign intercepts and 'Humint' dispatches reports that contained the "UFO" keyword. Humint was an Agency euphemism for human intelligence, actual spies, in other words, who occasionally happened to see some unusual things in the sky of the foreign countries where they did their work.

Next, he sent the same request via encrypted email to the NSA at Fort Meade, and to the office of the Secretary of the Air Force; he started tapping into his contacts within the academic community who were known to track anomalous phenomena, with an urgent request to assemble some data for him; and he prayed that something would turn up over the next seventy-two hours that would be good enough to satisfy Renslow and his powerful friends.

* * *

By the riverside, Peter Keller had all his senses on alert. He mistrusted Trent, and he didn't know what to think about the Asian woman. He spun around to find that the others had silently rejoined them. The bald fellow remained crouched down near the top of the bank, keeping watch.

"How did your "big stick" impress the Soviets, anyway?"

Junko didn't answer. She was just staring ahead at the water. It was the man in black who picked up the thread.

"Alintel showed them how the UFO material could be a source of immense energy, and make the balance of power go *tilt*. If we could do that just with the few pieces of junk we had picked up, can you imagine what we could do with a complete saucer?"

"Why are you telling me all this? Your buddies killed that poor pilot just because he saw a disk. So why are you so open with me about their existence now?"

"Because you're going to work for us, Mr. Keller," Junko told him, without the least emotion.

"What makes you so sure?"

"Exclusive rights to the entire story, Peter. Think about it."

"Wait . . . these fake saucers of yours . . . What's the point, if your outfit has gotten its hands on a real one?"

"A contingency plan, Mr. Keller. We need to understand how the revelation of an alien presence will impact society. Sooner or later the truth must come out. But we can't just tell the public everything and unleash a panic, can we? So we prepare people; we soften them up with simulations."

"You could have picked anybody," Peter observed.

"We need a damn good journalist, someone who can penetrate the cover-up, with a little help from us, of course. Remember Watergate? Where would Woodward have been without his 'Deep Throat' source?"

That last remark seemed to elicit merriment among the group, like an old memory of a good prank. Junko laughed. Even the man in black smiled.

"How do you know I won't just expose your little game?" snapped Keller.

They laughed even more.

"To reveal what, exactly? That most flying saucers are fakes?" asked Trent. "You haven't kept up with opinion polls lately, my friend. Four out of five Americans are ready to believe these things are extraterrestrial."

Junko picked up the thread: "We'd just call you a debunker," she said. "You'll read the snide remarks in all the New Age magazines. We'd ruin your show. 'Keller's part of the cover-up. He takes money from the CIA.' I can already see the ugly rumors."

"Probably beats up his girlfriends, too. No wonder he can't keep a decent job," chuckled the man in black.

"So what do you plan to do?" In spite of himself, Peter had to admit he was caught. And he was getting intrigued.

Again, it was the man in black who answered, suddenly serious, all business.

"We think the Kremlin shift in policy towards America had something to do with the captured UFO hardware. That's why we're interested in the man who appeared on your show—the one in shadow, your 'Mister X.' Can you tell us his real identity?"

Keller shrugged.

"Just some guy who's been sending anonymous tapes to the Brotherhood of Galactic Science. He claims to be a scientist. I've got no idea who he is. Or where he is."

Trent insisted, waving his hands in front of Peter's face.

"If he's seen the hardware, we need to know where it's being stored now, and what's being done with it. That could be the missing piece in the whole damn puzzle."

"So far all I've seen are your fake saucers," Peter argued. "How do I know Alintel has got a real one?"

"Lots of indications, no proof," Trent confessed.

"Is that why you picked me? Because I can ferret out some fresh information about your own bosses?"

"That's only part of the reason," Trent said. "If you can place the original tape into our hands, we might be able to identify your Mister X. Then it's up to you to make contact, to set up a meeting with him for us."

"What do you want from him?"

"We want to know where the base is. We want to get in."

"Well, once again, why me? If this story's so big, go to CBS, go to the *New York Times*."

"How do you know they're interested? Don't be naive, Pete. Who do you think runs those places? Besides, their reporters are too big, too visible. They'd never get within two miles of the facility."

Junko joined in, cold and precise, with the punchline.

"We'd be betrayed in a matter of hours if we went to the major media. No, we need someone who's slipped between the cracks. That's why you're perfect. Nobody cares much about you these days."

"Gee, thanks!"

Keller took a step back, sardonically chuckling and shaking his head as he looked from grim face to grim face. The wind whipped up Junko's hair. The Statue of Liberty loomed over her shoulder.

"Be our courier, Mr. Keller," she begged. "The American public has a right to know about the saucers. Alintel has been covering up too long. This is bigger than all of us."

Colonel Trent put it in plain business terms.

"Convince Mr. X to meet with us, and the story rights are yours, Peter. Yours exclusively."

"You people really push the right buttons, don't you?" asked Keller.

"That's what we do best," said the colonel, nodding his head in agreement.

CHAPTER ELEVEN

On Friday nights the offices of *Inside Story* were generally deserted, so Peter told Colonel Trent and Junko to wait for him in the hallway while he headed for the production area. He crossed the empty lobby, walked down the corridor, typed in his security code and plunged into the archives. Part of him was very frightened after the episode on Staten Island, but another part of him desperately wanted to delve deeper into the world his new contacts had just revealed to him. He turned on the lights and began rummaging through the videotapes.

It was like a game, the kind of intrigue that had first attracted him to journalism.

He found what he had been looking for, and right away he recognized that old fire within his mind, a passion for discovery he had not felt in years. When Ben Simon walked past the doorway he called him over urgently.

"So, you ready to get back to work after all?" his boss asked.

"Ben! I was hoping you'd still be here. I've got two of them outside, waiting in the hall!"

It was Ben Simon's turn to jump.

"The mystery men? Well, what're we waiting for?" He rushed outside.

But the hallway was empty. Keller looked around, alarmed and disappointed.

"I forgot—they're aliens, of course," Simon told him. "They're still here, but they've just turned invisible, right? In order to hide from us. Or did they just go through the wall?"

"Ben, I tell you they were here—just now!"

But his boss just sighed, his shoulders collapsing.

"This time I'm taking you off the show, Pete. Go tell your fucking story to *Fate Magazine*."

Keller frowned, shocked by his words. Ben went on: "You're not well; maybe you've been working too hard, harder than I've given you credit for."

"No, no, there's nothing the matter with me," Peter protested.

"Take a week off. Take two weeks off. You got at least that much coming in sick leave. I know it's not always easy, with little Sara, and your wife leaving you and stuff. But you'll make it. Everybody goes through things like that. You're a big boy. Get yourself together. We'll be here when you're ready to come back."

"I'm not crazy, Ben. I'm gonna crack this story, with or without you."

"Sure Pete, that's swell. Looking forward to it," Ben answered without conviction as he left. "I gotta get back to work. Rest up; we'll see you when you're feeling better."

Keller went away, tape in hand, and mad as hell. He barely looked up when the elevator stopped a few stories down.

The doors slid open.

Trent and Junko appeared out of nowhere and stepped inside.

"You shouldn't try and betray us, Pete. It's not a smart play," Trent said.

"Look who's talking about betrayal! The king of the mindfuck," Keller replied angrily.

But he didn't resist when Junko took the tape from his hands.

"You can't turn back now, Pete. You're in a whole new ballgame. From here on, trust no one—not even your own eyes."

* * *

Early on Saturday afternoon, with photographs and computer printouts spread over every available surface, the Alintel staff held a planning meeting. They had brought in a pizza and a few pitchers of beer, and they had already reviewed several stacks of documents. General Bushnell outlined his ideas and sorted out the possibilities.

A civilian, a simple fellow like Keller, would have found it astonishing that such a career military man could be at once so subtle and so reckless. Again, he selected young Greg as his target.

"Doctor Parker, please restate for us the goals of the present operation."

Greg knew he was in trouble whenever his chief called him "Doctor." He recited the mission's purpose as he understood it.

"To enhance the credibility of an extraterrestrial threat in the eyes of the world public; to discredit those who might guess that we do not have ultimate control over the UFO situation; and to flush out any competition that might interfere with our action from within the Intelligence community."

"Or from our scientific colleagues," added Beverly Bernard.

"The scientists don't count," said Joe Wilkinson, taking a long swallow from his beer mug. "Academia is a control system, too, but it's a closed one, and the rules are extremely strict. Their margin is nil."

"Don't you think that a good team of civilian amateurs, with the help of serious scientists, and some money, could do breakthrough research on UFOs?"

"They could do research, yeah, sure," said Wilkinson, wiping his mouth with the back of his hand. "Anybody can do research." He laughed: "It's a free country."

"You don't consider that a parameter?" Bushnell asked.

"No, because they'll never get close to solving the problem. There's no threat to us from that direction."

"Why not ?"

"Science is a good system of investigation, but it is designed to be directed at nature, a target assumed to not be biased or devious. Science often fails when one applies it to something artificial, or to a target that has an intelligence of its own, or a hidden agenda. Witness the mistakes that are often made in forensics."

"The UFO phenomenon must be part of nature," Beverly reminded him.

"Yes, but it is controlled by a non-human form of consciousness. *Ergo,* the study of it doesn't belong in science at all, it belongs in Intelligence. Meaning counter-espionage. Which is our own field of specialization."

"Many of the rules are the same," Parker pointed out.

"I disagree. In science, if you give me 95% of the data concerning a certain phenomenon, I'm happy because I now know 95% of what there is to know about that phenomenon, right?"

"What are you driving at?" broke in Beverly, growing impatient.

"Bear with me. In Intelligence, if I get 95% of the data, I have to assume that I have the cheapest part of the information. I still need the other five percent, don't you see?"

"But the price curve is different," she said, catching the point.

"Exactly—the price curve is different. I have to pay a much higher price for the next five percent. I may have to be willing to lose a few human lives to acquire it. Remember, Hitler had 95% of the information about the Normandy landing on D-Day. But he had the wrong 95%! Our side made sure of that. And he couldn't afford the last five percent."

"You're saying that all the statistics and databases the civilian groups are compiling can never get them close to the truth, then?" said Greg Parker.

Joe Wilkinson reached for his third slice of pizza and replied, "Even if they did, it would not get them any closer to our own objectives, which transcend understanding the phenomenon itself."

"Given the three goals you've summarized, where do we begin?" asked the general, taking control again, making sure all his pawns were at the right places on the chessboard. "If you were in my shoes, what course of action would you recommend? Come on, what would you tell Vulcan?"

"We must come up with a credible witness," Greg answered. "First we calibrate him; then we anchor him in our own version of the story, so befuddled by all the trash he uncovers that he'll never get to the actual truth."

"Do you think Keller's our boy? Will he take the bait?"

"Based on our psychological assessment, he's perfect," Beverly said. "He is cooperating with us already. He's entered the maze."

"Excellent. To quote T.S. Eliot, a wilderness of mirrors," observed Bushnell.

"I thought James Angleton said that," Beverly broke in, "when he was the legendary head of counter-intelligence at CIA."

"He said it, but he was quoting Eliot," the general reminded her.

* * *

At the other end of town, the Rogues were huddled around Junko and Trent in a nondescript basement room, studying dual video screens. On one of the monitors, they could see nothing but the obscured image of Mr. X telling his tale, the bizarre story Keller had already related to them. On the other, still frames of seemingly thousands of faces flashed by at breakneck speed.

Trent adjusted the controls on the console and gradually, steadily, the unknown man's altered voice became normalized. Further adjustments were made and the moving, fuzzy boxes over Mr. X's face became frozen as an image-processing program drew in contours that were then superimposed on the screen. As each piece of the puzzle became clearer, the computer sorted back and forth through the photos. Soon, a rough but discernible image of a face became visible; the screen locked onto the surprisingly sharp photograph of an angular, dark-haired man.

* * *

The sun was slowly going down over Westchester County on Sunday afternoon. Peter had gone back to his ex-wife's house, as he did every weekend, and he was pushing his giggling daughter on a swing set when Kathleen called to him from the back door:

"Phone call. It's for you."

As he went inside, Peter was irritated by his awareness of Gérard in the background, riding an exercycle in what used to be his study. He tried to ignore the fellow, and picked up the receiver.

"Hello?"

"About your friend Mister X . . ."

It was Colonel Trent's voice, all business.

"Why are you calling me here?" Peter almost choked, suddenly annoyed and scared. "How'd you get this number?"

The colonel ignored the question.

"The man's name is Ellis Denton, D-E-N-T-O-N, who lives in Sunspot, New Mexico. We'll give you one day to wrap up your affairs. You will fly to Las Cruces first thing Tuesday morning. Take a cab from there. Your ticket'll be in your mail box; expense money too. Don't forget to pack your video camera. And some blank tapes."

The line went dead before Peter could respond. He just stared at the receiver in his hand.

"Trouble?" asked his ex-wife.

"As always, right, Kat?" he remarked sadly, hanging up the instrument.

He went back into the yard, walked towards his daughter and knelt by the swing.

"I've gotta go away for a while, honey."

She could sense the sadness in his voice.

"Where you going Daddy?"

"Into a desert, far away."

"For how long?" asked Sara, sniffling.

He answered her by giving her a big hug, squeezing her tight. He didn't want to let go, as if he were holding her for the last time.

"I love you, Squid."

After a kiss on the cheek, he rose, stroking her hair and speaking softly.

"Be good. Mind your mom."

He glanced at Kathleen on the way out. She was at the kitchen door, watching him.

CHAPTER TWELVE

Early on Monday, Peter Keller took a cab to the financial district, entered a skyscraper a block away from Wall Street, and stepped into one of the many elevators to the higher floors. He walked up to a smartly-dressed receptionist who sat behind a large mahogany desk with three telephones. The sign above her read *Robertson, Benraph and Chernik, Esq.*

"Is Rachel Rand in?" Peter inquired, trying to keep his voice level and calm. "Can you please page her for me?"

Rachel walked in a few minutes later, and at first he didn't recognize the distraught woman he had met in Florida. She was dressed in an elegant white lace blouse, and a demure skirt of black wool that reached just below her knees. The surprise in her eyes was obvious as soon as she recognized him.

"Miss Rand, let me come straight to the point. I'm going to New Mexico to see a man who may be involved with the heart of the question we discussed the other day. I fly there tomorrow."

As he said it he realized how bizarre and crazy it all sounded.

"Why are you telling me this?"

"I . . . I was hoping . . . I want you to come with me," he finally stammered.

One look at Rachel told him she was flustered, caught off-guard.

A young attorney entered the room, law book in hand. He wore a three-piece pin-striped suit, and a red-and-yellow striped tie. He walked straight towards her, ignoring Peter.

"Rachel, I need those briefs ASAP," he said bluntly.

She stared at him, shocked by his rudeness. He must have realized that he was interrupting a serious conversation, because he corrected himself:

"Well, soon . . . okay?" Reluctantly, he took one step back, then: "Who's that fellow?" He asked, pointing at Peter, looking with disapproval at the blue jeans, the sneakers, the worn leather jacket.

"Lloyd Braun, meet Peter Keller. Uh, Pete's a television producer, from *Inside Story.*"

"Oh yeah?" The attorney almost spat the words in his face. "Well, your show's a piece of shit."

"So's your tie, buddy."

Peter turned his back to the man and spoke urgently to Rachel: "Will you come along? Please?"

She looked radiant and savvy, more beautiful than he remembered. At the time of the interview she had been tense, upset, under the stress of Borodine's interrogation, and angry at Peter's own blunders.

"Come where? What's he talking about?" asked the attorney. Rachel ignored him.

"I can't just pack up and go, Peter! New Mexico's a long way away."

At least she didn't just laugh at his offer, Keller thought. She was still looking at him, hesitating.

"New Mexico! Forget it, buddy, she's got way too much work to do right here in New York."

"Was I talking to you, Counselor?" Keller asked with heavy irony. He took a step towards the door.

Rachel was still standing there, biting her lip. She turned to face Braun.

"You'll find the briefs on the printer. The first draft is done, I'll proof it before noon."

He started to mumble something, backing away in apology, but she didn't give him time to find words:

"If you don't mind, it's time for my coffee break."

She tugged at Peter's sleeve. "Nobody's using the conference room. We can sit there and talk."

* * *

She was nervous when they sat down at one end of the long, glistening table of cultured marble, with the buildings on the other side of the avenue as a backdrop. The overcast had not lifted, and lights were on in all the offices; rainwater washed off the windows, blurring the scenes inside.

She poured him a cup of coffee from a silver-and-black container and smiled self-consciously, waiting for Peter to talk. He looked at her—the light brown hair falling easily on her shoulders, her upturned, mischievous nose, a sensuous mouth. Her face contrasted with the strength of her shoulders and her firm, precise gestures when she set the tray back on a sideboard, with a bit of impatience. Peter began.

"This whole experience, with the things—the objects in the sky ..."

He faltered, and began again: "I've given a lot of thought to what happened to you. I'm sorry I doubted you. When you see something like that, it changes your whole life, doesn't it? It's not just an intellectual thing. It plays with all your emotions."

He picked up the coffee cup and took it to his lips, thankful for the warmth and the rich flavor.

Rachel looked up and said, puzzled: "You're saying this as if you spoke from first-hand knowledge. You haven't shown much emotion before."

"I was fooling myself," he conceded. "In my business we're a bunch of cynics. We see people lying and faking all the time. It's easy to kill our own responses, to pretend we don't have human weaknesses. Because we're on the other side of the camera."

He could feel an urge to confide to her, to spill out his feelings, and he wondered if she, too, sensed a sexual charge, even at this moment of tension, when she was unsure of his sincerity.

"I've seen them," he blurted out. "What I've seen was only a faint approximation of the full-blown phenomenon. But I've met the people behind it. They're convinced the craft are real. They say there's even a real saucer right here on Earth, captured by the military."

"Great," she said unenthusiastically.

"Look at this," he went on, pulling out his wallet, extracting the telex message he had saved. "A pilot died in Nevada the other day, because he saw it come down. These people will stop at nothing to hide the truth."

"Oh, Peter," she said as she read the dispatch. Her voice shook. "Oh, my God."

"My contacts are fairly sure the solution to the whole puzzle is inside a hidden facility in the Southwest. Probably Nevada. They think they can help me get in there with a camera."

"Wait . . ." She raised her hand in the air. "You said you were going to see a man in New Mexico."

"He's supposed to know what's been going on inside that base. He can help us get back there. That's the only way I know to find out the real secret about what's been happening all those years, to you, to us—to everybody."

"When are you leaving?" she finally asked.

"JFK, tomorrow morning, Flight 27, American."

He turned around. She put a hand on his arm to stop him as he moved to get up. He took a deep breath, waiting for her to speak.

"I'll be there," she said. "But why is this so important to you?"

He noted how her hand trembled, how her lip quivered.

"I made you a promise once, and I broke it. I swore to make it up to you, didn't I? Also, now . . ."

"Now what?"

"Now I know you were telling the truth, Rachel. I have no doubt they exist any more."

* * *

It looked like any other bookstore in Greenwich Village, with dusty volumes in a narrow window, behind a glass pane nobody had bothered to wash in years. It dealt in occult and New Age topics.

Derisively, someone had sprayed the number 666 on the door, and the owner had not bothered to paint it over. Perhaps he felt that the dreaded Number of the Beast represented an additional attraction for his customers, most of whom were deeply in search of esoteric truths.

Peter Keller pushed open the door. He passed a pile of magazines on a low table, along with flyers of every color advertising lectures, tarot readings, and psychic consultations. Next came a cashier's booth cluttered with sticks of incense, candles and hanging crystals, display cases with statuettes and tapes, several containers of herbs and powders and many shelves of books.

There were books everywhere: on the tables, in piles on the floor, and behind a counter where a young woman in a long flowing dress was pouring a cup of herb tea for an elderly man who was writing notes into an equally ancient binder.

Peter felt disoriented among the pouches, the beads, the hieroglyphs and the bottles of dried leaves.

He asked, "Where would I find your section on UFOs?" and was rewarded by a quick side glance from the old man and a smile from the clerk, who waved her arm in a gesture that embraced the whole store.

"They're scattered pretty much everywhere," she said.

"Depends on what you're looking for, of course," the old man corrected her. He spoke in a gentle, helpful voice—a civilized voice, but there was a sharp edge to the tone, which Peter could not quite define.

"I'm trying to understand how the whole mess developed," Peter said, encouraged by his response. "Wasn't there an Air Force Project at one time?"

"You'll want to read Brad Steiger. He's in the paperback section," the woman said. "We have a video too, called *Cover-Up of the Gods,* which tells the whole story . . ."

"No, no, please don't listen to her," the elderly customer interrupted. "The new stuff is no good. Look among the used volumes. There may be something by Ruppelt, the officer in charge of the Blue Book project. Or good old Major Keyhoe; or Professor Hynek's book."

"Who was he?"

"Hynek? An astronomer in the Midwest. The Air Force hired him as their consultant. At the beginning he was quite a skeptic, even a debunker. Gradually he became a believer."

"You always make it sound like he told you everything he thought," the woman observed bitterly.

"As a matter of fact, he did," the customer replied. "I was on the faculty at Northwestern, teaching philosophy. He was director of the observatory. We used to argue endlessly about what those things were."

"My name is Keller," Peter said, stepping forward, holding out his hand. "Peter Keller."

"And I am Desmond Byrne," the old man responded. "Perhaps I can help you."

Slowly he made his way down the aisle, motioning for Peter to follow him. He pushed aside several fat bottles filled with dried leaves, bearing labels with long Latin names.

"Let's see," he mumbled, rummaging through battered volumes. "*The UFO Evidence,* a compilation by the NICAP group. You don't often see a copy of this."

"Is it any good?"

"As good as they come. An analysis of the best early reports, investigated by serious people."

"Could I contact them?"

"They don't exist anymore; their organization was disbanded years ago. This book is good too, *The Great Flying Saucer Hoax,* by the Lorenzens. They used to head up a research group in Arizona."

"Where are they now?"

"They're dead."

Peter couldn't help laughing.

"It sounds like all your buddies are either dead or more profitably occupied! So who could I talk to, to find out what's going on?"

The older man sighed, thrust a lurid pamphlet into his hand, and pointed at the flying saucer on the cover.

"If you're looking for quick, easy answers, there are several groups of believers like these," he said with irritation. "They mean well. You could join them for a few bucks. They'll send you a membership card that will confer upon you the title of 'scientific ufologist.' That should impress your landlady."

"I've already met a few of them. Is that really all there is about ufology?"

"Well, as a hobby it's not bad. It's less expensive than horses or women. Your mailbox will soon bulge with newsletters written by some pontificating ignoramus with a fondness for words like 'egregious' . . . You know what I mean."

Peter smiled. "That's not what I was asking. I want to know who's doing any serious research now. Anyone I could talk to?"

Byrne frowned. "Why on Earth do you want to get messed up in all this?" he asked abruptly, suddenly on guard.

"I am a journalist, preparing a documentary," Peter half-lied. Ben Simon would never authorize him to spend even five dollars on a used book by Ruppelt, Hynek, or anybody else. He was dealing with a younger public who didn't know anything about the history of the field, and didn't care.

"Are you on a deadline? How much time do you have?"

"I'm leaving town tomorrow morning," Peter said.

"Then forget all those books, and let's get out of here," Desmond Byrne said. "There's a coffeeshop just around the corner, where they serve the real stuff," he added, as he laid two dollars on the counter with a reproachful look at the clerk. "I can probably fill you in, if you can spare a couple of hours."

* * *

"The whole thing's turned into a circus nowadays. Tabloid shows stressing the more sensational elements have replaced the intelligent debate."

Professor Byrne was heating up again to a subject that had held his fascination for decades.

"Every time you turn on the idiot box you see people wearing strings of beads and crystals given to them by the kind 'space brothers,' and they keep yelling insults at other people who claim to have been implanted by evil aliens, all for the amusement of the populace . . ."

Then he realized that he was speaking to a television journalist, and he caught himself: "Forgive me, perhaps I am insulting your profession . . ."

"Go right ahead, Professor, be my guest. I happen to agree with you. What we turn out is mostly garbage, although we do stumble on an interesting story from time to time, and are able to follow it, provided we understand the background."

"Of course, of course," said Byrne, "and I promised to help you. Let's go back to our little briefing."

The coffeeshop smelled of Arabica and French blend and Brazilian roast, with a hint of cinnamon. They were seated in a corner, by the window. Two other customers, probably NYU students, were busy reading at separate tables.

"Have you ever heard of an organization called Alintel?" Peter asked hopefully. The man drew a blank.

"What are they supposed to be? A bunch of spooks?"

"Sort of . . ."

"There are so many of those! If that's what you want—paranoid theories—you can go back to that bookstore. You'll find all sorts of weird accusations."

"Have you ever heard that Reagan's Star War system, the SDI initiative, was actually directed against an alien invasion?"

"It's a very silly idea, Mr. Keller. If UFOs are real, surely they can evade our primitive particle guns."

"Then you don't believe in conspiracies at all, Desmond—is it all right to call you Desmond? You don't think the government knows?"

"Oh, the government knows, of course. They have always known there was a real phenomenon. But what could they do about it? It never seemed to be trying to invade us, or to attack us, so it could be safely forgotten, the data locked away."

"You do believe there's a cover-up, then?"

"In a way, yes. But this goes beyond a simple cover-up. You see, Peter, all the important decisions were made in the early fifties."

"That's when Blue Book got started? But you just told me Blue Book had been a cover, just a public relations exercise. Even Hynek . . ."

"Let's leave Hynek out of this, please. He was the only honest man in the whole bunch."

"Why didn't he expose the mess, then?"

"He was too careful. His reputation . . . He was a scientist in a narrow world of academia, on a campus in the Midwest. There were things he never knew."

"Such as . . . ?"

"The Air Force had initiated a massive study. A huge budget for the time. It was entrusted to the Battelle Memorial Institute. Then, in 1953, just as Battelle researchers were reaching their first conclusions, the CIA organized a panel . . ."

"The Robertson Panel," Peter jumped in, proud to display his recently acquired knowledge.

"You learn fast. The prestigious Robertson Panel, which was convened under Air Force cover, was tasked with an assessment of the phenomenon. Battelle was supposed to brief them . . ."

"Didn't they?"

"Oh yes—in part. They had to. But only after the leader of the Battelle project complained, in a secret letter, directly to his CIA bosses. He requested that the panel not be convened."

"Why not? He should have welcomed that panel. You're getting me very confused."

"He thought it was premature. He wanted to test more hypotheses. His staff had noticed heavy concentrations of unexplained UFO cases in specific geographic areas. He wanted to go test those areas by running simulations, in total secrecy."

Peter sat back in his chair and whistled.

"Fake UFOs . . . Very smart."

He thought of what the Rogues had shown him in the old stadium in New Jersey, of the craft that had hovered so realistically in front of him in the park.

"There was something else, an ominous little phrase in his memo, which has become known as the *Pentacle Memorandum.* That little phrase stated, in the strongest terms, that before testifying before the Robertson Panel, Battelle and their clients, meaning the CIA and the Air Force, *had to decide what could and could not be discussed at the meeting.*"

"More secrets, then."

"Actually, ufologists are divided on what these little words mean. Some believe that the Pentacle Memorandum is perfectly innocuous, a simple bureaucratic mix-up between a contractor and his client . . ."

"What do you think? What did Hynek think?"

"I told you to leave Hynek out of this. He was confused. We all were. Besides, as the great English cosmologist Fred Hoyle writes in his autobiography, astronomers seem to live in terror that someday they will discover something important."

"Didn't he have a copy of that document?"

"Sure, he had a copy of the Pentacle Memorandum, but it terrified him, so he kept it hidden away."

"Couldn't he confront the Battelle scientists?"

"He tried it once, many years later. They flew into a rage, snatched his notes out of his hands, tore them up and sent him home."

"Doesn't that prove they were still hiding something?"

"Not necessarily. They might have been embarrassed by the whole subject, wished they'd never heard of UFOs. Many scientists have felt that way. Heaven knows, I often feel that way myself."

Keller reclined in his chair, scanned the posters on the wall, and glanced without interest at the other customers. A new thought was forming in his head and, at the risk of antagonizing the old man, he had to try it out.

"The cover-up idea makes no sense at all, Desmond, when you come right down to it. I know that all the ufologists swear that the government is scared of releasing the truth because it might cause a panic, but it's a stupid notion."

Keller drained his coffee and stared furiously at the bottom of his cup as if he were about to read the future of the galaxy in it. Peter continued.

"I'm a journalist, damn it. Let me tell you, as a professional, there are a thousand ways to release that kind of information gradually, so ordinary folks won't get scared of it, or feel cornered. If those so-called aliens are trying to attack us, and they have all this super-technology, why is it taking them half a century to invade a third-class planet like ours?"

"Of course, if you put it that way, it's a stupid notion," said the professor. "The danger of a public scare is minimal. But there is a real cover-up, Peter."

"What's the point of it?"

"It isn't the reality of this particular phenomenon they are covering up, it's what they've done with it: All the little horrors they can't reveal, the sins they've hidden from us all these years. If they came clean with the UFO data, a lot of dirty government machinations would come out with it."

There was a long pause. Keller shuffled his feet, and finally asked: "More coffee?"

"Why not?" the old man replied. "At my age, there are few pleasures left."

* * *

Peter carried the two cups over to the counter for a refill. He came back with a tray, and two slices of cake. He noticed that the students had left. Now Desmond and Peter were the only two customers.

"What do you really believe, Desmond?"

"Belief is not a scientific term."

"Don't you ever speculate?"

"I suppose I could, as long as you label it clearly as such." Suddenly the old man was all business. "I think that 1953 was the turning point. Ufology became a cover for mind-control research. Forget the Robertson Panel. They only reached the conclusion they did based on a few selected cases, biased data, and an aborted briefing from Battelle. They noted that the phenomenon was of marginal interest to science, not a threat to the nation, and should be debunked. They said the civilian UFO groups should be infiltrated and watched. That was predictable. They said there were no UFOs. What else could they say? Project Blue Book continued to serve as a lightning rod. They used Hynek. It was a public-relations job, as I've already told you. Behind the scene . . ."

He gestured towards the crowd passing on the sidewalk, the fellow who sold hot dogs at the corner, the kids going home after class.

"Look at all these people," the old man said. "They obey unquestioned control systems; they get up when the sun rises, they go to sleep when darkness comes. They cross the street when the light is green, they freeze in their tracks when it's red. They lead lives of utter drudgery—but they do vote, and tomorrow, if some politician tells them to go off to a faraway land with a loaded gun and die there for some obscure cause, they will do it. What is driving them is fear. Like animals you see at the circus. Or rats in a maze."

It was the second time in as many days that Keller had heard a human crowd compared to a bunch of animals—"the flock," Colonel Trent had said.

The old man went on, with contempt in every word.

"They fear everything: a God they've never seen, their own wives, and also the leaders they've elected, the cops whose salaries they keep paying. They even fear themselves. They fear their own creativity, they stifle their own aspirations to be anything but fearful little beasts."

Fairly stunned by this outburst, Peter felt glad they were the only customers left. He would have been embarrassed if they had been overheard. Yet, in an odd sort of way, the old philosophy professor was stirring thoughts that were his own.

"So what actually happened to ufology in 1953?"

"A bifurcation. The Air Force went on overtly with a perfectly innocent project called Blue Book, which was not supposed to ever find anything. And the Intelligence community created a covert, underground effort aimed at exploiting the popular belief in UFOs and at duplicating their effects on human subjects."

Peter sighed. "You lost me again," he said.

The old man patiently tried to explain.

"UFOs, whatever they are, can paralyze people, and stop engines at a distance. If you were a military agency, wouldn't you want to understand how that's done? That would mean conducting experiments with everything from simple flashing lights to pulsed microwaves."

"You mean, like radar beams?"

"If you like, but over a lower range of intensity—otherwise you would simply cook your subject from the inside out. The effect of pulsed microwaves on the human brain is most varied and interesting, Peter. So is the effect of extremely low frequency, at the other end of the electromagnetic spectrum. You can induce hallucinations, make people hear words, change their mood, perhaps even evoke visual images. But you won't find much of this data in the open literature."

"You ramble, Desmond," said Peter. "What does that have to do with flying saucers?"

"Maybe nothing. I hope to God it has nothing to do with it. I would like to believe, as most ufologists do these days, that we are simply being visited by warm fuzzies with big black eyes who come from Epsilon Eridani."

"But you don't think that's the answer."

"No, damn it! I don't think it fits the situation. I think we're being shamelessly manipulated."

"By whom?"

"Suppose you could make a lot of people believe that we were about to be invaded by extraterrestrials. It's a simple idea. Anybody can grasp it. You'd start with the elite, the intellectuals, the artists. Gradually you'd have the media spreading the imagery ... People like you, Peter."

"It wouldn't work. The media doesn't have that much power. Basically, we're just entertainers. The scientists would always remain skeptical."

The man shrugged, dismissing the notion.

"Who cares about them? Scientists don't shape public opinion. They've contradicted themselves too many times, they confuse everybody. In a world of mutual assured destruction, where war has become impossible, something has to replace weapon development. Something has to keep the economy going, the people in line, the populace in its place. They have to fear something, right? Otherwise you'll end up with ungovernable chaos. Look at Bosnia. Look at Africa. Look at what's left of the USSR, for heaven's sake!"

"So you would engineer a whole new source of fear. Is that what you are saying? Fear of aliens from the end of the galaxy?"

"Exactly. Our Gods always come and go in the sky, don't they? The dragons of China, the Pillar of Fire, the Virgin Mary. Even Jesus. In every culture. Except that in this century the old gods have trouble impressing the techno-wizards. They're not very credible. They can't just send us the old dreams and visions. Burning bushes don't fool anybody any more. We need to build a whole new Pantheon. That's where E.T. comes in."

"And those guys, the geniuses within the Intelligence community, you believe they thought of that back in 1953?"

"Why not? If you can manufacture an H-bomb, or a microwave beam that can stop tanks and confuse brains, surely you can also manufacture the appropriate new religion to go with your new toys."

Keller got up. He needed some fresh air, and felt his head was about to explode.

"What if the UFOs really exist? Nuts-and-bolts machines? What if the government had actually captured some hardware?"

"There is no proof of any of that."

"What if they did?"

"Then it would be even more horrible than anything I have told you. My speculations would be mere ramblings, the babbling of a child."

"You would believe in conspiracies, then?"

"Hiding a real UFO would demand the ultimate conspiracy."

"And the prize?"

"The ultimate prize, Peter. Planetary control. But of course . . ."

"Of course what?"

"To catch the prize, eventually they would have to produce the evidence. Or get somebody to produce it for them."

It was a phrase Peter would ponder often in the next few days.

* * *

As soon as he came home Peter Keller turned on the television set, hoping to catch the latest installment of *Inside Story.* To his surprise, it had been preempted by a series of grim reports from the former Soviet Union, where civil wars were threatening, from North Korea which was again threatening to use its nuclear weapons, and from Mexico, where ancient quarrels between the government and minority populace, fueled by the ambitions of various internal factions, had precipitated riots and massacres.

"We regret to announce that tonight's edition of *Inside Story* will be preempted," the anchorman said, between two commentaries from the field over a background of gunfire and general bloodbath.

"Something has to keep the populace in place," Desmond had observed only two hours earlier. It had sounded like the delusion of a harmless old intellectual. "Human beings have to have something to fear."

Yes, he had to concede that the belief in UFOs could be used to create such fear. But where was the proof?

"What kind of journalist am I?" Keller suddenly thought. He hadn't even asked the professor if he had any of those old documents he kept quoting, like that *Pentacle Memorandum* he had referred to. He picked up the jacket he had thrown over an armchair, and he made his way back to the Village.

It would be easy to find Desmond again. The old man had said he lived in the house across the street from the coffee shop. A lot of people seemed to have clustered around that particular intersection by the time Peter reached it, however, all looking in the same direction.

When Keller caught up with the crowd he smelled the smoke and the soot and he saw the flashing red lights of the firemen and the ambulances. There was a shape on a stretcher, with a sheet over its head. Peter gradually realized who must be lying under it.

"They say it was a gas leak that did it."

The owner of the coffeeshop had recognized Keller. "Was he a friend of yours?" he asked with concern.

"I only met him today. We talked about some old books."

"He loved to talk about old books," the man said. "I bet he died happy."

CHAPTER THIRTEEN

When the dusty taxicab from Las Cruces Airport left them at the base of the main building, Peter and Rachel saw the smooth tower of the solar observatory at Sunspot rise above them—a tall, spiraling structure, chalk-white and futuristic. Situated on a mountain peak, it overlooked the beautiful desert setting of New Mexico, made even more spectacular by the brilliant, clear-blue sky above.

The landscape reminded Peter of a fine summer, a lifetime ago, it seemed, when he had traveled across the Southwest with Kathleen and little Sara. (How old was she then? Four? Five?) He remembered his daughter's wonder at the majesty of the Grand Canyon, their happiness during the days when they traveled through the high desert, pitching their tent in the evening, and the awesome beauty and immensity of the night.

Now he was back in the bright savagery of the sand and the dust, his mind burning with uncertainty, driven by an obsessive desire to uncover the ugly secrets of other men.

They went into the building, awed by the clean, vertical architecture that carried the telescope two hundred feet beyond the top of the mountain. Two repairmen wearing goggles were working high up in the rafters, precariously perched on metal girders around the large instrument. They were replacing a section of the mirror assembly that dominated the center of the structure.

A staff member, a young woman in a bright dress, got up from behind the desk where she had been typing. "Can I help you?" she enquired.

Peter gave her his business card, and the name of the man he was looking for.

"We don't have an Ellis Denton here," she told them. "Wait, I'll ask Stan and Frank. They know the town pretty well."

The two men stopped their work and looked down when she called them from the ground below. One of them made a helpless gesture.

"Can't hear you, Samantha! We'll be down in a little while."

They turned back to their unfinished task, and while they adjusted the delicate instrument with the fine moves of a surgeon, the young woman invited the visitors to wait in a small auditorium, its walls covered with huge photographs of the sun, blazing gloriously against the blackness of space.

Peter and Rachel were drinking their second cup of coffee of the day and chatting with the receptionist when the two technicians entered the room. They offered an interesting contrast. Frank was the younger one, about forty-five, blond and bearded. His companion was older, smaller, wiry, balding. They both wore white overalls, and the young one had a baseball-style workcap.

"We've been all over town, and no one seems to know him," Rachel was telling Samantha.

"Who? What's going on?" asked the older technician, the one who was introduced as Stan.

"These folks are in from the East Coast. You ever hear of an Ellis Denton?" the woman asked.

Stan shrugged and shook his head. Frank seemed to be searching his memory.

"Didn't think so," Samantha concluded. "Sunspot was created just for solar study, you know. Everyone who lives here has something to do with the observatory. We'd know Ellis Denton if there were anyone by that name around here."

But Frank seemed to have recalled something.

"You know, there is a guy named Benton, or Denson, out towards White Sands. Sort of a hermit character. Is that who you're looking for?"

"Could you tell us how to find him?" asked Keller.

"I can do better than that," Frank said, wiping his hands on his overalls. "I'll take you there myself."

* * *

They were following a winding mountain highway, and Frank's pickup truck was rumbling along, its flatbed filled with tools and camping equipment. Across the back window was a gun rack holding two rifles. As the vehicle twisted and veered, Peter became increasingly aware of his leg pressed against Rachel's thigh, while their driver lit a cigarette and turned on the radio. A song came on over the hum of the tires and the roar of the engine. It was "Riders on the Storm."

"All the way to Sunspot to see a guy you've never met before?" Frank said suddenly. He seemed to find the situation very funny. In fact, he laughed heartily, perhaps a little nervously, at the very thought.

"No one ever, ever comes to Sunspot unless they have to. What's so important about Ellis Denton?"

"That's personal," Keller responded, a bit bluntly.

"We really appreciate your help, Frank," Rachel stepped in, sensing the mounting tension. "Taking time off work to help us."

"That's not much of a problem," the man growled. He added gruffly, "any extra money I make, the government confiscates it anyway."

She glanced up at Peter, raising an eyebrow as if to ask, "What does the taxman have to do with this?"

"Just the same, it's nice to see folks in Sunspot are friendly," she persisted with a smile, trying to cool down the situation.

The truck took a corner a little fast, the tires screeching in complaint. Rolling thunder was crackling in The Doors' song.

"I'm not being friendly," Frank said between his teeth.

The pickup was gaining speed, going much too fast for the treacherous terrain. Tires wailed as the vehicle blasted through a hairpin curve, barely staying on the road.

Rachel looked at Peter again and saw that he was as alarmed as she by the repairman's reckless driving. Frank was becoming a changed man—trembling, nervous, his eyes on fire.

"I was at a bar in Crownpoint last week," he said, practically spitting the words out. "Some cowboy sits right down next to me, starts askin' where I'm from. The guy didn't fool me. He was a little too clean for a cowboy, see? Just another spook come to eyeball me, to intimidate me. Like you two."

Peter and Rachel looked at each other, confused by the sudden violence in the man's voice.

"I've seen you all come and go," he asserted bitterly. "That journalist who pretended he was a free-lance writer for *Omni*, and the hitch-hiker, and the traveling salesman ... and the young fellow with the sharp-looking glasses who carried a card from the Institute for UFO Analysis ... *Investigator*, it read, real impressive. But I could see half an inch of thin wire sticking out of his shirt. They should train them better. Do they think I'm completely stupid? And the girl ... the one I trusted."

A pause, then he added somberly: "I stopped trusting her when I caught her staring at me, asking too many questions."

His eyes went back to the road for the briefest moment, just in time to take another hair-raising curve on two wheels.

"She said I was paranoid," he added with a nervous laugh.

"Gee, I can't imagine why, Ellis," said Keller.

Rachel jumped:

"Ellis?? You mean ... he's Ellis Denton?"

Instead of responding, the irritated technician sped up even more and radically passed a slower vehicle. The pickup fishtailed, almost spinning out, as it took another turn. Keller kept his cool.

"I've come with an offer," he said.

* * *

Under George Bush part of the West Wing of the White House had been redesigned, turning some of the formal reception areas into office space, and it was in one of the new conference rooms that President Henry Templeton and his Chief of Staff Bill Renslow were scheduled to attend a special briefing by the CIA's Keeper of the Weird, Dr. James O'Grady.

In spite of his repeated statements warning the White House that the Administration was going off on a wild goose-chase, Presidential Science Adviser Dr. Brent Holborn had been told that his presence was expected. As a long-term rationalist, he thought the meeting was a waste of time, and he attended under duress. He understood the circumstances, however, and feared that political convenience would once again dictate technology policy. Influential friends of the president and of the First Lady had insisted that the UFO issue must be scrutinized. That had happened before. In this instance it was a major financial backer who had personally recommended that the President take notice.

So it was, on that gloomy Tuesday afternoon in autumn, that Dr. James O'Grady drove over from CIA Headquarters with a set of transparencies and samples of various official regulations stored in the black briefcase that rested on the front seat of his Ford Taurus, which he drove carefully through the Key Bridge traffic. While the radio played a soft classical tune, he reviewed in his mind the essential points he would make before the chief executive. His reflections were interrupted by an urgent radio bulletin.

The Mexican situation, which had been quietly heating up to a sizzling urgency, had suddenly turned ugly, the newsman said. A coup was clearly in the works. Tanks had been seen converging on Los Pinos, the palace of the president. And masses of refugees were reported heading for the U.S. border at Ciudad Juarez and in Southern California.

He pushed the urgent news out of his mind. His colleagues at the Central America desk would be the ones to lose sleep over that situation. He continued down Pennsylvania Avenue, took the side street as he had been instructed, and was escorted to the conference room as soon as he showed his credentials to the guards.

Left alone in the room, O'Grady had time to organize his documents in neat piles on the large table. There was a stack of CIA computer listings, mostly abstracts culled from the world press, mentioning UFO sightings of every shape and form. He had also brought two intercepts of foreign telephone conversations on the subject, and samples of the recent writings of some of the more rational UFO believers.

Having reviewed his papers, O'Grady checked the projector, put his first slide on the glass plate, adjusted the focus and sat down to wait.

Brent Holborn came in first, a quiet, affable man with thin white hair, a goatee and steel-rimmed glasses. He introduced himself politely, clearly unimpressed by the young CIA analyst.

"I think this meeting will be a waste of your time and mine," he observed with a sigh as he sat down, placing a thin manila folder on the table in front of him. "I don't mind telling you that was the gist of my report to the president. But he and the First Lady have influential friends who believe the UFO issue should be scrutinized."

"Bill Renslow told me as much when he called me from the White House. Do you happen to know where the pressure comes from?" asked O'Grady.

"The chairman of an agribusiness concern in Topeka, Kansas, I think. Wealthy fellow. Close friend of the senator who heads up Defense Appropriations, no less. He was a contributor in the campaign, which carries a lot of weight. In my current role as director of the Office of Science and Technology Policy, I can't ignore these things." He gave a brief chuckle. "The 'policy' is the tail that wags the 'science' dog, in this case."

President Templeton arrived with Bill Renslow before O'Grady could ask any more questions, and the four of them sat down.

"You look like a fine young man, Dr. O'Grady," Templeton began, "but from what I hear, a large sector of the population has become convinced that your agency is hiding the truth about the reality of unidentified flying objects."

"Our mission is to keep official secrets, Mr. President. We are often criticized because of it."

"Keeping secrets from the general public is one thing; keeping them from me is quite another."

Brent Holborn protested. "Those who argue that way are only a handful of hard-core believers, Mr. President," he said before O'Grady could formulate an answer. "They have been saying it since 1947. There's no actual proof. Never was."

"Let him answer, please," insisted Templeton. "Did you bring any of the classified stuff today?"

"Yes, sir, I did ... but some of the unclassified data is equally valuable."

"Let's begin with the secret information, if you don't mind."

"Very well," said O'Grady, who had been forewarned about the president's romantic affinity for titillating intrigue. "These listings are transcripts of foreign conversations about UFOs that we have acquired in the last sixty days. As you know, our listening posts . . ."

"I don't see anything very convincing here," said Holborn, quickly scanning the thin computer printout. He sighed loudly. "I suppose the only reason it's classified stems from the intercepting technology, right? Because we don't want friendly governments to know just how extensively we eavesdrop?"

"That's right, sir," confessed O'Grady. "The reports by themselves are primarily about lights in the sky over France and Belgium."

"Carl Sagan would be able to explain away ten such cases before breakfast any day of the week," said the science adviser, rubbing it in. "What else do you have in your bag of tricks?"

The young analyst kept his cool.

"My predecessors at the Agency have left me a thick file of photographs, including a remarkable series taken in Florida, which our confidential civilian sources have analyzed."

The president himself rose from his chair to turn down the lights, and the four men scrutinized the screen in silence for a few minutes. Again, Holborn was the first one to speak.

"Are those classified, too?" he barked.

"Yes, sir."

"But I recognize some of them from a rather popular book that was published a few years ago. How can you classify something that's in a book anyone can buy at the corner drugstore?"

The president shuffled his feet.

"The source is classified, sir, because it is the analysis that counts, in this case," interjected O'Grady.

"What did your source conclude?"

"That the pictures were indeed genuine."

"Have you checked it yourself? They look fake to me."

"I did not have the resources to go to the site, Dr. Holborn."

"Then you shouldn't be talking about something you can't substantiate yourself. Not in this setting."

O'Grady looked around him at the paintings on the walls, the American flag, the bust of Abraham Lincoln. He swallowed hard, and didn't answer.

"Let's skip the photographs," proposed Bill Renslow. "We're

wasting the president's time. What else do you have?"

"I have prepared a summary of the history of the subject."

"Let's hear it," the president said with a sigh, turning the lights back on. O'Grady blinked in the sudden illumination.

"It all began in the summer of 1947. The Air Force is said to have retrieved some unidentified hardware from the desert of New Mexico, and on the 23rd of September the Commander of the Air Materiel Command wrote to the chief of Air Intelligence Requirements, stating that the phenomenon was real and not illusionary."

Brent Holborn had opened the manila folder he had brought into the room. He extracted a sheet of closely-typed text and raised his hand, again interrupting the CIA analyst.

"I have here the full text of the letter by the officer in question, Lieutenant General Nathan Twining . . ."

"With all due respect, sir, what's the relevance . . ." interrupted O'Grady.

"It's very relevant," said Holborn. "Twining talks about UFOs in general being a real phenomenon, as you said, but he never addressed any alleged crash data. You are taking his comments entirely out of context. In fact, his letter specifically goes on to deplore, and I quote, '*the lack of physical evidence in the shape of crash-recovered exhibits*.' So if something crashed in New Mexico, he certainly didn't know about it."

"Be that as it may," objected O'Grady, "most ufologists believe there has been a compartmentalized study of extraterrestrial debris all along."

"Again," said Holborn, "I hate to be such a pest, but the CIA—your own employers, young man—issued a secret briefing paper on August 14, 1952, five years after the supposed crash, stating, and I quote, '*there is no shred of evidence to support*' the spaceship theory."

O'Grady was about to respond when a Navy lieutenant came into the room, snapped to attention and approached the president, placing a piece of paper in front of him.

"Ask them to get in touch with the National Security Council immediately," Templeton told the lieutenant, adding, "I'll be along in a few minutes."

"I am afraid we have to cut this interesting discussion short, gentlemen," he told Holborn and the others as he rose from the table. "I have enjoyed our brief talk, although we seem to be a long way from having any sort of hard data about this issue. Dr. O'Grady, if you happen to come across interesting developments in the future, I hope you will communicate them to us through Bill Renslow's office. There is a tense situation developing in Ukraine, as you know, on top of our troubles in Mexico and the Middle East. If Moscow and Kiev start resorting to nuclear weapons, we may never have the luxury to speculate about intelligent life-forms in outer space."

"Perhaps people will finally wake up to the fact that we live on a very small and very fragile planet," said Brent Holborn.

"The Bible warns us about sowing the wind, and reaping the whirlwind," mused Renslow.

The quotation seemed to trigger something in the president's mind, because he turned around, took a step back towards them and smiled:

"One of my predecessors was very concerned with Biblical prophecies, as you know," he reminded them, pointing to the framed photograph of a pensive Ronald Reagan in the portrait gallery that adorned the back wall of the conference room. "Perhaps *fascinated* would be a better word than *concerned.* He made repeated references to the fact that if we were to learn of an imminent attack by interplanetary forces, we and the Russians would soon be united in a common effort to save the planet."

"He seemed especially worried about eschatology," the science adviser agreed.

"What is that?" asked Templeton. "One of those ancient segments that didn't get into the official version of the Bible?"

"No, Mr. President. *Eschatology* comes from the Greek *Eskhatos,* which means 'last,' and *logos,* which means 'discourse.' Eschatology is the science of the ultimate ends of humanity, of that which follows terrestrial life. And the Eschaton, I suppose, means the end of the world."

"I take it you don't really believe in all that, do you, Dr. Holborn?"

"No, sir, most scientists don't believe the universe will end very soon, although we do expect the sun to burn up eventually, and to destroy the solar system as it does so, in a few hundreds of millions of years."

The remark brought a wan smile to the president's lips.

"My Administration will have completed its work by then, gentlemen, although sometimes things move so slowly in Washington that I have some doubts. In the meantime, there are enough voices clamoring about UFOs to warrant a response, to reassure the public, if nothing else. Why don't you see what you can do quietly along those lines, Dr. Holborn? We could assemble a White House Task Force, as we did for aging or health reform. This may even be your opportunity to become as well-known to American television viewers as your colleague Dr. Sagan, whom you mentioned earlier."

There was a pique of irony in the president's words. The science adviser knew high-level politics too well to miss it.

"Let me tell you a little story I heard from one of my predecessors in this job, Jack Gibbons, who served under Bill Clinton. When people pointed out to him what a great opportunity he had, after a rather dull life as a solar physicist, to become a Cabinet member, he reminded them of the tale of the old man and the frog."

"The old man and the frog?" asked the president. "I don't think I have heard that one."

"This old man was walking along quietly, minding his own business, when he saw a little frog on the path. 'I am not really a frog,' said the frog, 'I am a beautiful princess. And if you would but kiss me, I would resume my human form, and we would lead a happy new life together.' He looked at her—"

"—I would have kissed that frog," the president said. "Any politician would have kissed that frog."

"Well, the old man didn't. He picked her up and put her in his pocket. Soon he heard her protesting from deep inside his jacket. 'You don't understand, sir, I don't think you have heard me, I am not a real frog but a fine princess, who would be your youthful lover if you would but kiss me.' He took her out of his pocket and said: 'I did hear you, my dear, but at my age, and under the present circumstances, I would much rather own a talking frog.' And that is how I feel about the UFO business, Mr. President."

CHAPTER FOURTEEN

The sun was high over the bleakly beautiful landscape, highlighting the dunes of the White Sands Missile Range in the distance below. Peter and Rachel had reached Ellis's cabin, perched up in the mountains, and their guide was boiling something in a complex, homemade solar stove. Rachel sat nearby, and Keller leaned on the shady side of a rock, swatting at mosquitoes. He had just told Denton everything he knew about the Alintel Rogues.

"Needless to say, they're quite anxious to meet with you."

"Yeah, sure," said Denton mockingly, "and they sent you over here to hand-deliver their kind invitation? Is that it?"

"They expect me to twist your arm, if necessary."

Ellis ladled some leafy-looking food onto their plates. Rachel started eating it tentatively, but Keller grimaced as he saw that it looked like a cross between spinach and squid. He put his plate aside, and swatted another mosquito.

"Can't we eat lunch indoors?" he asked irritably.

"My home's bugged. The observatory too. So we eat right here, or we don't eat at all."

Ellis ate a spoonful of food, leaned back and looked straight at Peter.

"So, Mr. Keller, it's clear you make a fine errand boy. But what's in it for you?"

"A mother of a story—if you come through. And that's a big IF, if you ask me. How could a handyman like you have seen something so important?"

"Don't let these working clothes fool you. I hold a doctorate in high-energy physics from Cal Tech. Five years at Sandia Labs, two more at CERN in Geneva, on the particle detectors for the LHC."

"What's that?"

"The Large Hadron Collider. Never mind what it does. Supposed to be operational in 2002. And then . . . then I took that fucked-up SDI job in Nevada."

"Why did you take it, Ellis?" asked Rachel.

"What a challenge—*Star Wars!*" He turned his head away in disgust. "But there was another project, a hidden agenda behind Star Wars. Most of that money was siphoned off to the side, going for a completely different purpose."

"And nobody knew?"

"Nobody was interested. Oh, we did have congressional staffers show up from time to time in their fancy limousines. They brought their girlfriends, spent the night in wild parties and were too tired in the morning to pay much attention to anything we told them. The science was way over their heads anyway. It was just a big boondoggle. So we'd hide all the real stuff, and arrange for a big show of laser blasts to wow their little brains. And the CIA guys, what a joke! You'd see them coming from a mile away."

"Are you saying that you were keeping the project secret not only from Congress but from our own Intelligence folks?"

"Does that surprise you? We were under the cover of a black project, remember, using a private corporate structure, spending mostly private money. All they could ever see was the top level—whatever we decided to show them."

For a moment, Keller was at a loss for words, his mind spinning, remembering the man in black's speech on the train. Gradually, he managed to put it in perspective.

"Could be more paranoid crap; could be just another crank theory," he told Rachel.

"Hey, some of my ideas may be a little out there, but this isn't a theory, pal! I was there! I know what was going on." The man sighed, then laughed. "Wait a minute here! Is that how you plan to entice me to come down this mountain to meet these 'Rogues' of yours? By challenging my credentials, by insulting me?"

"We can't break this story without them," Keller said. "It's as simple as that."

"Who says I'm even interested in breaking the story with you?"

"It's your best way of staying alive. Perhaps your only chance," Peter added, thinking of the old man he had met in the bookstore in Greenwich Village. "Assuming what you say is true ..."

"It is true."

"Then you can't hide up here forever."

"Why not? I'm not afraid of the government. I've got my guns, and I know how to use them. Nobody's interested in this story anyway."

"Give me some proof, and I'll make sure your story reaches the folks out there. A plain statement, on prime time. We have an audience of forty million. Alintel could never touch you after that."

As he thoughtfully, worriedly considered Keller's offer, Ellis took a bite of his food. Peter did the same, grimacing.

"Uggh ... this is awful."

"It's macrobiotic. Probably the only truly safe meal you've had in ages. You've been eating crap so long you can't taste real food."

"On the contrary, I *can* taste it. That's the problem. I should introduce you to my friend Gérard," Peter said bitterly.

"What do you mean by 'safe'?" Rachel asked.

"The powers-that-be are putting something in the food chain, something that pacifies the public."

"Oh sure. They put it in the fluoride too, I suppose," Peter stated sarcastically. But the joke was lost on Denton.

"By the way ... I've been wondering ..." He turned to Rachel. "Where did you first see my tape, anyway?"

"It's Peter who saw it, at the headquarters of the Brotherhood of Galactic Science. I've been a member there until I—" Rachel began. Ellis cut her off.

"You've been in The Brotherhood? Both of you?" He asked, suddenly in awe of them. "Have you ... Have you actually met Dr. Borodine?"

"We've both had that immense honor," Peter answered drily.

"I've been trying to contact Borodine for two years now, but all I get back in the mail are requests for money. I find him a fascinating man."

"Well ..." Peter began. But Ellis was too excited to listen to his doubts.

"You know what Dr. Borodine's problem is? He's got the wrong message. If I could just convince him that there are different kinds of aliens, I could get my message to the millions who follow him. And now that they're landing ..."

"Whoa! Now that WHO is landing?"

"Haven't you seen his latest show? He made the big announcement yesterday. The aliens spoke to Borodine; they're supposed to be landing tomorrow night at the Four Corners Festival!"

Excitedly, he added: "I'm planning to be there, of course, but I doubt I'll get within a hundred yards of him ... Unless you could get me in to see him. If you think you could. It would mean a lot to me."

"Oh, so now you want something from us! Well, Dr. Denton ... maybe we can strike a deal after all," Peter said, suddenly amused.

* * *

By mid-afternoon they had reached the New Mexico highlands, on a road that cruised over stark, lunar terrain. Again, it was Denton who was at the wheel, driving fast but not as recklessly as he had that morning. Beside him, Rachel and Peter were listening to his recollections of the black project.

"Yeah, the whole experience screwed me up pretty good," he was saying. "I saw something back there I wasn't supposed to, and I've been looking over my shoulder ever since."

He removed his baseball cap, to show them his hair.

"You think I like this look? You think I like living life disguised as a blond?"

Keller laughed.

"My heart bleeds, Denton. If you want peace of mind, you've gotta put some proof in my hands."

"Don't fret—you get me face-to-face with Borodine, and I'll get you close to the so-called Star Wars test site."

"Not good enough," Peter said, brandishing his camcorder. "I want to see this underground base of yours. I want to film inside."

"You'd never make it. It's high security—ground sensors for miles around."

"But you got in," Rachel observed.

"Yeah. Not sure why, though ... Seemed like an accident at the time," he said thoughtfully. "Of course, I don't believe in accidents anymore," he grinned.

"While the sun is still up, I want to do an interview of you, just to get it on record before we get close to the base," Peter said, pressing his case. "Please. In an hour or so there won't be enough light. We can stop near those trees, and I'll get you on tape, Ellis."

"No, let's drive on," Denton replied, "we're still a long way away. I don't know about you, but I'm anxious to get to the Four Corners in time for the landing. I want to see Borodine."

Peter and Rachel fell silent, sensing that Ellis was anxious to change the subject. Abruptly, he turned on the radio.

"The roads towards the borders are crowded with refugees from the fighting around Mexico City," said a local newscaster. "President Templeton has reaffirmed that the United States would only intervene if the situation threatened our citizens and ..."

"They wouldn't be fighting like that if they knew what's really going on," Ellis observed, turning off the set. "I'm sick and tired of all that hypocrisy. Next thing you know, they'll be sending the fucking UN down there."

"Why don't you let us tell the story, then? You promised to make a statement for us, on-camera," Keller reminded him.

He shrugged impatiently. "All right, all right, Mr. Big Reporter. I'll make a statement."

"Can we stop somewhere, then?"

"I don't have to stop to talk to you. You can shoot while I'm driving."

"It won't look very professional," Keller complained. "Especially with all the bumps on the road, the way you're barreling down the highway."

"I don't care. We're not in Hollywood."

Reluctantly, Peter leaned against the right-hand door, wedging his body against the dashboard, to get a better angle on Denton's face.

* * *

"Suppose we start with a little background. How did you first become interested in UFOs?"

"I didn't," Denton stated forcefully. "No, really, I didn't have any interest in UFOs, at first. I just happened to get involved. It was an accident. I'm a physicist, as I told you. I never even read a UFO book until I was hired by a contractor under the Strategic Defense Initiative. They gave me a special clearance. I had to take a whole battery of tests."

"What kind of tests?"

"Psychological. They wanted to make sure I was stable enough to go through the briefings and still retain my sanity."

Peter and Rachel exchanged a brief look. Ellis's sanity was not completely obvious to either one of them.

"Who briefed you?"

"Corporate types."

"Not military?"

"Like I told you, corporate types. Don't interrupt me all the time, or this won't go anywhere. We had to sign special non-disclosure papers. It seemed routine at the time. We were a small part of a very big project that went all the way back to the fifties, when the military was secretly collecting and analyzing foreign technology data."

"What kind of data?"

"Before Sputnik, it was all planes and bombs, radioactive fallout samples, ocean technology. Later, rockets, spent boosters, and actual satellites."

"The name of the project?"

"It went through many names, as you would expect from studies extending over several decades. At the time of the Robertson Panel the Battelle UFO group was under Project Stork. It was still in existence under the name White Stork in 1956, and even in '57 you find traces of it, in connection with analyses of hard traces left by UFO landings, turf samples, that kind of stuff. Even in the mid-sixties . . ."

"We'll be checking all this," Peter interjected.

"Be my guest," Ellis snickered. Of course, if you ask about Stork or White Stork, you'll be told it was all innocuous reverse engineering of Soviet aircraft and the like."

"So those Battelle guys were involved in the documents that the

Air Force has released, like the famous Report 14?"

"Yeah, but principally in Report 13, the top secret one, which has never come out. People are told Battelle skipped that number because the Pentagon was superstitious."

"Cute," said Keller.

"Report 14 was a smokescreen. The conclusion section is nonsense, even though the text itself is fine. You have to look for what's missing: any discussion of technical instrumentation that could be used to study UFOs, for one. Any mention of what could be done at the time with our own high-performance aircraft and radar. And you have to look at the outcome."

"You mean, after Robertson?"

"Yeah, just look at the CIA. Walter B. Smith set up the panel just before he was replaced as director. Follow the timing here, it's important: the panel met in January 1953. Exactly three months later the notorious order that instituted MK-ULTRA was signed. Responsibility for UFO analysis was taken away from the Air Technical Information Center. Over at the Air Force, Captain Ruppelt was out. Everything went black."

"Who got the assignment?"

"The Air Intelligence Service Squadrons. They were located all over the country. They did the secret fieldwork. They didn't have to deal with the press, the CIA, or anybody else. They could keep things away from Blue Book and Hynek, who never saw the good data. They funneled their best information to whatever group was set up under the cover of the Robertson Panel debunking. It must have gone to those 'Rogues' of yours at Alintel. They fit the bill pretty well, from what you've told me. But I never crossed paths with them."

"Whatever happened to Blue Book?"

"It was terminated in the late sixties, after a two-year study at the University of Colorado that was a mockery of science. Has anybody told you about the Bollender Memorandum?"

"No. All that stuff's new to me."

"You should look it up. General Carroll Bollender was with the Air Force. He wrote it in October, 1969, unclassified. But there are sixteen attachments, mind you, and now they 'cannot be found.'

This memo was the basis for the cancellation of Blue Book. It makes it clear that the project was never on the distribution list for any UFO reports that could affect national security, which 'continue to be handled through the standard Air Force procedures designed for this purpose.' It's hard to be more clear. Blue Book was a public-relations job all along, and they didn't need it any more."

"And what about the Pentacle Memorandum?" asked Keller.

He thought they would leave the road when Denton reacted so violently that the steering wheel seemed to jump in his hands.

"Where did you hear about that?" he asked.

"Never mind. What do you know about it?"

"It was written by a scientist at Battelle, Howard Cross, who warned the Air Force that it was the wrong time to set up the Robertson Panel."

"Yet they went ahead and ignored his advice," Keller pointed out.

"Maybe not. They had an urgent need to throw a cold blanket over the whole subject, especially since they went into a mode of simulating close encounters to gauge people's psychological reactions."

"How did they do that?" asked Peter, who was beginning to see a pattern.

"For one, they controlled the cults. Ever wonder why so many strange cults suddenly popped up in the mid-seventies? Remember the Peoples' Temple?"

"Yeah, what about them? You're drifting into paranoia again, Ellis. Jim Jones was a nut. They couldn't have controlled him."

"Are you so sure? Did you know he once claimed to have lived in the constellation Orion? He also told his disciples he had met a mysterious individual once, in a parking lot. This is the man who taught him how to control people. And what was he doing meeting with the head of state of the island nation of Grenada, the same man who demanded that the United Nations investigate UFOs?"

"You're not saying that Alintel engineered the massacre in Guiana?"

"Maybe not. We'll never find out, will we? Ed and Jeanie Mills, two of Jones's assistants who had left the cult, were assassinated in Oakland before they could fully testify. Before they died, they repeatedly warned that there would never be any serious investigation of Jim Jones, because of his cozy contacts at very high political levels. They were too scared to talk."

"That could be just your theorizing again. How did they die?"

"In their own house in Oakland one evening. The assassin killed Ed first, with a 22-caliber bullet in the head, then he tracked Jeanie down to the bathroom and executed her. Typical handiwork of a professional assassin. Yet the police claims that it was their son who did it."

"So? Maybe they're right. What does the kid say to that?"

"He swears he was watching television in the next room and didn't hear anything."

"What about the weapon?"

"They never did find it."

"Ellis, that doesn't prove anything. It's just another theory. You think many cults are controlled that way?"

"Yes."

"For what purpose?"

"Multiple benefits. These cults are a great cover. Nobody suspects a nut of being a clever spy. An intelligence agency investigating UFOs or other borderline phenomena can gain tremendous leverage from such groups—both negatively, as smokescreens to muddy up the waters, and positively as information dragnets, or as limited hangouts. Also, a religious organization is an ideal setup for money-laundering."

"Let's assume you're right. Why did it have to be so deeply secret, going back to the fifties?"

"Because by then the strategic implications were already mind-boggling, and they knew it. That's where our own project originated."

"Were you hired under White Stork, then?"

"No, I went in under a successor project. Battelle was no longer involved. At first I was assigned to instrumenting the SCIF ..."

"What do you mean by SCIF?"

"The base, our laboratory," Denton answered impatiently. "SCIF stands for 'Special Compartmentalized Intelligence Facility.' Which means access to the base requires a special clearance. Things became complicated real soon. There were rumors about the capture of actual hardware, actual alien artifacts. One night, when we were relaxing away from the lab, we noticed a lot of activity around the base, and the next day we were told we couldn't go there anymore, that the SCIF had been fully activated."

"Yet you got back in."

"Once. Only once. It was in the afternoon. One of the managers came and called a dozen of us together. They herded us into a bus with blacked-out windows, like I said on that tape you watched."

"Couldn't you see anything?"

"No, there was a curtain between us and the driver, so we couldn't even see out the windshield. We left the staging area where some of the maintenance labs were, and they drove us to a brand new base a few miles from there. I know the approximate direction. I could estimate the distance from the speed of the bus and the bumps in the road."

"What did you see?"

"You won't believe me if I tell you. It was horrifying. There were various objects there, parts of a UFO maybe, but it certainly didn't match what people imagine when they talk about flying saucers."

"How was it different, Ellis?" Peter's voice took on an edge of excitement.

"It's impossible to describe. It's nuts and bolts, all right, yet . . . everything the public has been led to believe, all the speculation in the media, and the reports to Congress and the president and the CIA—all that is a crock of shit. They had no choice but to lock it all up and throw away the key. That project has a life of its own now, and it has cut itself off from the rest of humanity. That's all I will tell you. The rest of it, you'll have to see for yourself."

"Come on Ellis, tell us the truth!"

"What makes you think you're ready for the truth?"

The anger and the bitterness of the morning had returned. Denton was agitated, his hands squeezing the wheel to hide his trembling.

"I've tried to tell the truth. God knows I've tried. Even paid with my own money to make that tape you saw. I've tried to leak the truth out a thousand ways. Nobody ever took it seriously. People would rather go on believing whatever their preconceptions tell them. They'd rather think I'm crazy."

"We're different, Ellis. We didn't come here with any preconceptions."

"That's what everybody says," he spat out. "You want to know what truth does to you? It doesn't make you free, that's for sure. It just makes you alone."

Ellis kept driving along the dusty road, his jaws clenched, and Peter thought he was about to cry. Rachel gave him a pointed look that meant it was time to turn off his camera if he didn't want to antagonize their eccentric guide. He took the hint. Yet in the back of his mind the mystery had expanded again, and his eagerness to shake the truth out of Ellis Denton had only grown. So he sat back against the bench seat of the pickup truck, feeling furious, and stared ahead at the coming darkness.

CHAPTER FIFTEEN

The sun had set by the time they reached the Four Corners area, and they joined a large gathering of the faithful under the bank of floodlights that marked the entrance to the Brotherhood's convention site. The makeshift parking lot was already filled with cars; hundreds of people milled about before a large elevated stage, like concert-goers awaiting an open-air rock show. The three travelers left the pickup where attendants dressed in white jumpsuits pointed to an empty space, and joined the line of the faithful who were waiting to buy tickets for the event.

On stage a mysterious shape, several stories high, stood concealed by a giant tarp. Large video screens flanked it, showing a steady stream of BGS commercials, appeals for money, and excerpts from Borodine's past lectures.

Occasionally, the screens would simply be filled with the BGS logo, and a spotlight would hit Brother Ralph, who addressed the gathering crowd from a stand-up microphone at the front of the stage.

"Welcome, Ambassadors," he began, speaking into the public address system. He had traded his three-piece suit for a purple robe trimmed with gold. "Yes, I did say *Ambassadors*. For tonight, we are all delegates from the human race."

The ticket-taker was a nerdish youth with an attitude, wearing a T-shirt that depicted a landing saucer beneath the words CONTACT! I WAS THERE. On the back was an image of Borodine looking up in anticipation. The disciples were flocking past, having paid thirty dollars each. Keller noticed a BGS member selling the same T-shirt in a variety of colors. Reluctantly, he handed over ninety dollars to the smiling ticket-taker.

"Where does the money go?" he asked. "Does Borodine need a new radar?"

The well-trained ticket-taker ignored the question without allowing his plastic smile to waver for one second. He handed each of them a packet containing an admission stub, a pocket flashlight, a toothbrush, a picture of the Milky Way with a "You are here" arrow pointing at Earth, and a bumper sticker saying *Warning! I Brake For Extra-Terrestrials.*"

"Excuse me," Rachel asked a woman who was walking by, a wire pyramid over her head, "but what's the toothbrush for?"

"Oh, that's very simple, really," she answered in a shrill voice. She pulled out a similar implement suspended around her neck on a purple string. "Suppose you're abducted by the aliens tonight," she explained. "Wouldn't you be embarrassed if you ended up on board a spaceship, hundreds of light-years from Earth, without your toothbrush?"

"You've got to admit she has a point," Peter remarked as they walked past a concession area where BGS members hawked crystals, shirts, sweatsuits, jackets, mini-pyramids, herbal teas and plastic flying-saucer models.

The crowd was a strange mix. People with flowing white robes and purple stars on their foreheads mingled with senior citizens in golfing slacks who had just gotten off the bus from Iowa. Peter noted that the T-shirts were selling briskly. Rachel, in fact, went to stand in line to get one.

"My boss was right," he said to Ellis Denton. "What a racket!"

"First time I ever heard you say Ben was right," observed a cheerful voice near him.

Peter spun around, surprised and delighted to see Thomas Archer standing a few feet away.

"What the hell are you doing here?" he cried out, hugging his friend. Thomas laughed at his surprise.

"Just following up on the good doctor. If the Brothers are gonna land, we figured we might as well get it on tape."

Ellis had been eyeing Thomas suspiciously.

"Who's this, another spook?" He asked with a frown that chilled the congenial atmosphere.

"You wanna run that by me one more time?" asked Thomas, anger suddenly flashing in his eyes.

"Easy, man, he just thinks you're some kind of spy," said Keller with a shrug. "Meet Ellis Denton. He says that to everybody."

Ellis nodded and moved off without further comment, heading toward the stage.

"Nice company you're keeping, Pete. What gives? I thought you were lounging around Manhattan, taking it easy. At least now I know why you haven't returned my calls."

Before Peter could reply, Thomas' attention was diverted to Rachel, who was returning with a BGS shirt. He was surprised to see her move toward Keller.

"Well, hello! It's Rachel, isn't it? You guys here together?"

Peter hesitated. It was Rachel who answered:

"Yes. We're together," she said matter-of-factly. "Anything wrong with that?" She looked at him with defiance in her eyes, taking Peter by the hand.

Thomas grinned, shot his friend a knowing look, eyebrows raised.

"Sick leave, huh? You feelin' better yet?"

"I'm working on a story," Peter said.

"Uh-huh," mumbled Thomas with a smile.

"She's been abducted—she's dealt with this stuff first-hand. I haven't. It's as simple as that."

Peter was earnest, but neither Thomas nor Rachel seemed to be buying it, so he quickly changed the subject.

"You're going to shoot an interview with the good doctor, aren't you, Tommy? Then I need a really big favor for my friend Ellis . . ."

* * *

It was a nightclub like many others in the Washington area—just a seedy room with a long stage, curtains around the walls, three bad musicians, a couple of projectors that punched holes in the darkness, and some twenty tables occupied by tired bureaucrats and visiting Japanese businessmen. There was a smell of stale beer and cold cigarettes, mixed with the girls' heady perfume.

General Bushnell liked the place because it was within a couple of blocks of the office he used on K street when he was in town—

a convenient cover. Greg Parker and Joe Wilkinson had come over with him, and they were in the mood to celebrate. Yet they were hardly looking at the shapely dancer who was taking off her black stockings on the stage, or at the two cocktail waitresses, a tall redhead and a statuesque blonde, who were standing next to the bar, plainly bored out of their minds.

"He's late again," Greg noted with annoyance.

"He'll be here," Bushnell assured him. "Let's get some drinks."

It was the blonde who came over to their table and took their order, dressed in an improbable costume two sizes too small for her bust.

"Doesn't that hurt?" asked Greg.

She nodded in agreement.

"I can't wait to take it off," she replied, bending over to be heard over the loud music from the band, her breasts two inches from his face. "I even know a place we can go to if you want to help me remove it," she suggested.

"What about my friends here?" he teased her, pointing to Bushnell and to Joe Wilkinson. "They get very lonely when I leave them."

"No problem," she went on, catching the eye of the redhead. "Karen can take care of them."

"We're waiting for another guy," the general said. "For now we just need something to drink."

"And some pretzels or something."

Bushnell motioned towards the door.

"Here he comes now."

Greg turned around in time to see Colonel Trent lifting a corner of the red velvet curtain that closed off the entrance, keeping the cold air of the street away from the cozy intimacy of the club. He wore beige trousers, a navy-blue blazer, and a striped tie. He looked around, saw them, and walked over to their table.

"Why did you want to meet here?" Trent asked as he sat down. "There's got to be twenty secure places within a mile of this dive where we could've talked."

"There's so much noise here, this is as safe a room as any for what we need to discuss tonight," Bushnell observed. "As long as we keep the terminology vague enough."

"Besides, it's the general's favorite evening spot," added Joe Wilkinson.

"Why don't you just tell the young lady what you're drinking, so we can get started? We don't have all night."

She noted their drink orders and swiveled away from their table, all four men following the motion of her hips.

"We could've met at the Boeing place over at the Beltway," the colonel insisted.

Bushnell shook his head.

"I didn't want anybody to see you join us. You're supposed to be a Rogue, Trent—or have you forgotten your assignment?"

"How much longer do we have to keep up the charade?"

"We could keep it up indefinitely. Keller'll never catch on. He's too pleased with his adventure as an intrepid reporter; figures he's the next Woodward."

"Are we so sure that he won't catch on?" asked Parker. "Something could go wrong."

"Not as long as Trent does his job properly. But remember, if you step on your dick, I've never heard of you."

"Still, we oughtta be careful with Keller," Joe Wilkinson said.

"I'm not worried about him," the colonel stated with finality, crossing his legs. "It's Ellis Denton who bothers me. Why did we send Keller chasing after him, instead of using one of our limited hang-outs?"

"Like who?"

"Like the Institute for UFO Analysis. They would have done a better job."

"What's wrong with using Denton?" asked Bushnell.

Joe butted in. "The guy's totally paranoid at this point. He's unpredictable. Carries an arsenal wherever he goes. Could get out of hand. Now he's taking Keller to the Four Corners, to talk to Borodine."

"So what? The plan is moving along smoothly. What are you concerned about? Beverly's over there, keeping an eye on the good professor."

"Denton will lead them to the Base; he knows where . . ." Trent stopped in mid-sentence, his eyes following something behind them.

This time it was the redhead who was walking over to their table. She sat on the closest empty chair, perched on the edge of the seat.

"Denise tells me you fellas are looking for some fun," she cooed, crossing her stockinged legs so that the tip of her right shoe came to touch Bushnell's calf, massaging it softly. "Care to buy me a drink?"

"Good idea," assented the general, resting a hand on her knee. "But I haven't talked to my buddies in a long time, and we need to catch up."

"Well, be that way," she said, getting up with a shrug that sent her hair tumbling over a bare shoulder. She cupped her bulging breasts in her hands, looked him straight in the eye and added, "when you're ready to catch up on some of these, let me know, will ya, big boy?"

"Where were we?" asked Trent when she had gone back to her corner.

"You were worried about the base," Joe said.

"Not really. We've got that situation under control. It's the Fast-walker that bothers me."

"Welcome to the club," grumbled the general. "I was hoping you were bringing us some good news from the scientists."

"I wish I could," the colonel answered somberly. "Problem is, the thing's still not responding the way it should. We've gone back to the old hypothesis, using magneto-hydro-dynamics as the primary propulsion model, and it doesn't fit the data at all. So we're back to square one."

* * *

Professor Borodine traveled everywhere in a luxurious three-axle coach specially equipped for long-range expeditions. Bristling with antennas and fully air-conditioned, it was painted to resemble a spaceship, and even sported a satellite dish and a small radar intended for a yacht. Inside the bus, in Borodine's high-tech living-room on wheels, Thomas and his camera crew were preparing to interview the leader of the Brotherhood. An eager-eyed Ellis Denton sat in a corner with the *Inside Story* crew, looking on with interest and pretending to be part of the team.

Ellis half-smiled to himself. If Borodine could accept him as one of the technicians, perhaps he would grant him a private interview afterwards. And Ellis would finally be able to share his enthusiasm and his frustration with someone who would read into his mind and understand his experiences.

Borodine spoke to them, his eyes on a mirror, checking the shape of his best profile. He was in a remarkably good mood, relaxed and happy, with a superior grin on his face.

"Lots of folks saw that show you did on us. Brought us some new parishioners. Some people say you tried to ridicule me. If you did, it didn't work. Tonight, I get the last laugh."

He reclined in the plastic-smelling luxury of his flagship, surrounded with the expensive toys that proclaimed the pinnacle of success he had reached. A graceful assistant, a young woman who wore a blue toga and not much else, applied some makeup to his face, busied herself with his hair and fussed with his eyebrows.

The prepping done, Borodine spun his swivel chair around, faced the camera and loudly clapped his hands together.

"So turn that damn thing on. Let's get some truth out there to the American people."

* * *

"What do we care if we understand the physics or not?" asked Greg. "It doesn't really matter how these things fly."

"Trust a sociologist to say something like that," said Joe contemptuously.

Parker looked vexed. "Well, I mean, it would be *nice* to know how they fly. But for our own purposes in the next few critical days, that doesn't really matter, does it? We'll have plenty of time afterwards to sort out those measurements of yours."

"It does matter to us, as scientists," Wilkinson retorted. "It's not enough to know that they're real."

Greg took a swallow of his drink. "That's where you and I disagree. If it's real, then it has a politics. That's enough reason for us to get fully involved."

"But will it influence the planet in the way we intend, if we can't make them fly at will? Under our own control? With our own pilots?" Joe asked.

"In terms of influence, it doesn't really matter whether they're real or not, whatever 'they' may turn out to be physically. That's what you fellows don't see," the general stated. "I tend to agree with Greg. What matters is that many people—not just a few crazy ufologists but influential people, folks with mansions and private jet helicopters and bank accounts and billion-dollar stock portfolios—now realize that the phenomenon impinges on their existence, on their *business,* on their ability to manage the planet. And they have started acting on that belief."

"But that's not really new," the colonel broke in. "That battle used to be fought in terms of faith. Christian dogma. Many people equate flying saucers with the New Age, and the New Age with the Antichrist."

Greg Parker was becoming angry.

"Again, you're not listening. Whether it's Christ doing it, or the Great Whore of Babylon, or the Short Grays from Zeta Reticuli, as the ufologists keep saying, I really don't give a shit. I have no problem with any of it being true, or all of it being true at the same time, that's not our job."

"What's our job, then, as you see it?" Trent asked.

Greg Parker leaned forward, resting his elbows on the table.

"Our job is to set the reference standard for the control level," he asserted. "The control of beliefs. In that sense, UFOs are just another geopolitical problem, like Korea was in the fifties, or Nicaragua, or any of the strategic hotspots on the globe. We're simply dealing with competing management systems . . ."

". . . as governments always do," noted the general, completing the sentence.

* * *

"Where does the Agency, I mean . . . the funny folks across the river, where do they come into all this?" asked Trent. "Because I'd rather not have the fucking CIA in my way over the next few days, if I can help it."

The general grabbed a palmful of pretzels and laughed.

"Just be thankful we're the only two agencies that are involved—Alintel on one side and the CIA on the other. They don't even know we exist. You don't need to worry about them. To begin with, their constituency is quite different from ours. And of course, given the methodology they're trained to use, they can't deal with the category of issues we handle. That's why Alintel was created so far away from the Intelligence Community mainstream: We could keep a secret of this magnitude indefinitely, and they couldn't. Their analysts are essentially intellectuals, and this problem doesn't have anything to do with intellect. It doesn't call for subtle analysis any more than that girl's cleavage over there."

"It's a question of weirdness," Greg broke in. "Remember how Hynek invented a strangeness scale? We're so far along his strangeness scale that no ufologist will ever find us, no matter how many times they picket the White House or clamor about a cover-up, and neither will the CIA. They keep looking for us in the wrong places."

"Still, the Agency has assets in the field ... I think you underestimate them," Trent went on, frowning. The general was unimpressed.

"All their assets are known. They are visible, they stick out of the landscape like the Washington Monument, or the control tower over at Dulles. Everybody knows who they are and what they do. Everybody knows they own the UFO groups. And a lot of good it does them. In the meantime we monitor their channels and feed them whatever'll keep them happy. Right now they're deep into psycho-social models, instigating surveys of religious attitudes, studying the new religious movements, infiltrating pagan groups, all that fun stuff; that's the new fashion over there at Langley. They get a lot of free oil massages, but they haven't figured out the real problem yet, after all these years."

"But they have clearances up the gazoo," Trent insisted.

"The folks with the clearances are the easiest ones to fool," Bushnell said forcefully. "If you work them from the inside."

"How come?"

"Two reasons. First and foremost, the simple arrogance of privilege. They think they know everything, because they are privy to a multitude of esoteric details that the man in the street doesn't have access to. Second, we know exactly who they are. One clever look in the computer, and we can find out their backgrounds, the nature and limits of their clearances, the date of their last security check, whether they're gay or straight, what they know and what they don't know."

"How do you propose to fool them?" Greg asked.

"Easy. Simple geometry. You just look for a *locus,* the place where their ignorance profile intersects their arrogance."

"But they have a burning interest in UFOs, whatever they are," Colonel Trent asserted. "They can and will exploit the phenomenon, just as we do. You've got to give them credit for that."

Bushnell nodded.

"Sure they can. They already have exploited it. Look at all the fun they have at Gulf Breeze, planting their little stories. They're extending the precepts of the Pentacle Memorandum, taking the work of Battelle to its logical conclusions, using the saucers to gain inspiration for their non-lethal platforms. You've got to hand it to them—they're consistent folks."

"But they don't have the real hardware."

"No, they don't have the real hardware, and they're still wasting their time on counter-intelligence exercises, on various distractions, the Roswell Crash and the Aviary and the MJ-12 conspiracy and the rest of that pleasantly warm shit we splattered on their windshield."

The four men laughed, then fell quiet as a new girl came on stage.

She climbed a curved staircase to balconies at a height of six feet or so, and started undressing to the strains of "Goodnight, my Love." She was a brunette glamour queen, all effervescence. She gradually unveiled well-exercised legs and a flat stomach, sliding her few remaining garments over her shoulders as she sang in a hoarse, sexy voice.

For a moment they forgot that there was a base somewhere in the desert, where the world's most closely guarded secret was hidden away.

* * *

It was Joe Wilkinson who brought their thoughts back to the purpose of the meeting.

"Leaving aside what we do, the day-in, day-out stuff, what's the fundamental question now? What are the priorities?"

"Well, you know as much, or as little as I do about the hardware," replied Trent. "You're aware of our painful failures, the depth of the unknown we're staring at. The real problem is the transition. What the believers perceive as the Eschaton is really a planetary transition. That's what we need to manage."

"There are psychiatrists over at Harvard who are saying the same thing, and all the abductionists too. They're making a run at planetary consciousness," Greg Parker pointed out.

"Exactly, and you know where the money comes from, for all that hypnosis work," said Trent. He grinned. "Anybody can play that game. Anybody with a nine-figure bank account, which means a lot of folks these days."

Joe Wilkinson cut in: "You don't even need that much money. If you own a few magazines, or a few TV stations ..."

"The new priest-kings and their off-planet friends ..." mused Greg. "Yes, those are the players. Some day it'll make a fine doctoral dissertation at Berkeley or Harvard."

"Don't bet on it. One fact is certain: We're making a transition on this planet."

Parker acknowledged it, then looked at Trent, and said, "The real message of the hardware we've captured is that we may be slaves."

"Or food," the colonel pointed out.

"You're wrong. You're both wrong," the general said brightly. "We're making a transition from being monkey-people to being something else. That's the real message. We're managing an Apocalypse."

"But there isn't a single 'something else,' there are many," objected Trent. "And every end-point, every alternative, has a destabilizing potential for somebody: an oil company, General Motors, Japan, Inc., who knows?"

The blonde came back to their table carrying a tray loaded with glasses.

"Are you gonna spend some time with us or not?" she asked. " 'Cause if you're just sitting here talking philosophy all evening, me and Karen will go find some live ones somewhere else."

"We'll let you know later, honey," Trent said, brushing her off. "We haven't even started drinking seriously yet."

She stuck her tongue out to him, left the glasses on the table and walked away. When she was out of earshot the colonel went on in a confidential tone.

"The point is, we've got the hardware, and we're the only game in town. We can produce it any time we want, as we've shown the Ruskies. That's a destabilizing factor that we control."

"Forgive me," said the general, "I grew up in Oregon. I don't understand all that sociological jargon you guys use all the time, but I do understand trees and forests. When you want to fell a large redwood, you need two things. One is a big chainsaw, of course, but that's not enough. You also need to decide where the tree is going to fall. That's where the crew spends most of its time, did you know that? You don't want it hitting other trees, d'ya know why?"

"Minimize scarring?" Greg guessed.

"That's the answer you'd give the ecology nuts, of course. But the real reason is that you want to make sure it falls in one piece. If it breaks you lose some of the value. It's a waste of money, pure and simple."

"What's your point ?" asked Colonel Trent.

"Simply this: The ability to suddenly unveil the alien problem in the world media, that is our chainsaw." The general motioned up and down in the air as if he held a heavy cutting tool. "We can use it overnight, any time we choose. We can get on the front page of the *New York Times*. But the key to the Eschaton, the transformation of civilization as we know it, lies in a single decision. Before Vulcan goes in to brief the president, we need to decide where the tree will fall. And it had better fall in one piece."

"Preferably on his head," concluded Joe Wilkinson, to the laughter of the group.

"How much does Henry Templeton know about all this?" asked Greg.

"Almost nothing. He's just asked the CIA for a briefing. They have one fellow there, Jim O'Grady, who's supposed to be the Keeper of the Weird. He gets all the stuff nobody wants: poltergeists and sea monsters, witches and spoon-bending, Bigfoot sightings and crop circles—he's got closets full of them."

"Should we worry about him?" inquired Greg.

"Hell, no, he's over in left field, snooping on a few ufologists, scrutinizing their bank records, tapping phones—the usual bureaucratic routine, kid stuff. I told you, their constituency is different from ours. O'Grady sits over there in his office at Langley. He's got all those impressive clearances, he thinks he's a superspy, he's sure he knows everything. His reality is defined by the four or five different badges that dangle around his neck. He doesn't even suspect we exist, in his wildest dreams."

"He will, as soon as we widen the circle of our initiations," Joe pointed out. "The guy's no dummy."

"Naturally," General Bushnell said on a tone of finality. "We'll bring him into our little scheme when we're ready. He has a role to play, like everybody else. That's the nice thing about the end of the world—everybody gets a piece of the action."

The colonel looked at his watch and got up.

"Listen, I'd better fly back to New Mexico," he said, "my friend Pete will need me pretty soon."

"Your instructions regarding the handling of Keller will be on the computer net," Bushnell said. "In the usual code."

"Remember what the general said," Wilkinson pointed out, only half-joking. "Don't step on your dick."

Greg and Joe pushed back their chairs and got up too.

"We were planning to have dinner around here somewhere," said the bearded man, leaving his glass unfinished. "Care to join us, General?"

"No, I think I'll stay here a while and enjoy myself," Bushnell said, looking over towards the blonde. "Tomorrow's gonna be a long, long day."

CHAPTER SIXTEEN

The show had begun with fanfare and ceremony, and the kind of flamboyance the Brotherhood was so adept at deploying. Soon Peter and Rachel found themselves in the middle of an expectant flock of enthusiastic New-Agers, illuminated by banks of intense mercury lights. The speeches began. The crowd was buzzing expectantly when Brother Ralph spoke over the public address system, and pointed to the sky.

"Stay alert out there, friends. Our Visitors are on their way through the cosmos. They won't be long now."

The crowd cheered. The wind blew from the desert, bringing a dryness that made people cough, and putting a strange exultation into their blood.

"And now, ladies and gentlemen, we'd like to acknowledge a very special and wealthy guest who is here with us tonight. Please welcome world-famous playboy, entrepreneur and television personality, Mister Nik Tok!"

As the BGS band played "Money Makes the World Go Round," the self-made celebrity walked across the stage, decked out in his trademark bright-white suit. He sported his characteristically huge grin and enthusiastically shook Ralph's hand before taking the mike.

"On behalf of Nik Tok Enterprises, I present BGS this check for ten thousand dollar!"

He turned to the cheering crowd. "I do seminah this Sunday night in Albuquerque, Albuquerque Civic Center. I invite all you peoples . . ." —he pointed up to the sky— "and all Space Brother to join me and learn how to enjoy life more better!"

He pumped Ralph's hand again for good measure, and exited the stage, waving to the appreciative audience. Lost among the believers, Peter was shaking his head.

Thomas made his way through the crowd and sidled up to Keller. "I left your wacko friend Ellis in the fancy bus with the wacko Doctor. Last I saw they were yakking behind closed doors, about advanced propulsion technology and space-age shit."

The entire crowd suddenly began to cheer. Rachel joined Peter and Thomas, and together they looked to the stage to discover the cause of the general exhilaration.

"Ladies and Gentlemen, friends of Galactic Science and of the Space Brothers everywhere—the Doctor is in!" shouted Brother Ralph.

The crowd went wild; the stirring strains of "Also Spracht Zarathustra" pierced the night. A spotlight hit center stage, revealing Doctor Borodine in all his glory. Kettle drums pounded away, and the crowd gasped as he appeared to levitate straight up into the air, courtesy of a hidden condor crane. He came to a stop fifty feet above the ground, his voice booming out over the loudspeakers:

"And I heard them in my head: *Tonight, Helmut,* they said . . . *tonight, give us a signal, a beacon—for tonight we land!*"

His words reverberated across the desert through a dozen loudspeakers.

"And so it is with great pleasure that I unveil our latest and greatest achievement in Galactic Science—" his voice quavered. "The brand-new Integratron!"

On cue, the tarps covering the huge shape on stage were pulled away. There was a momentary awestruck silence, followed by thunderous applause from all directions.

The Integratron loomed three stories high. The top part was a metal ball, thirty feet in diameter, at the tip of a large crystalline column that towered into the sky. Along the base of the structure, huge concave dishes faced the crowd. To the side of the stage, a ring of blue landing lights illuminated the partially completed Solid Gold Landing Strip.

"Upon my command, the Integratron will collect all the positive energy emanating from you tonight, and modulate it into a coherent beam that will reach to the heavens, thus lighting the way for our Brothers from the Stars. Count down with me, and watch the skies, everyone! Watch the skies."

The cheering crowd screamed: "Ten, nine, eight, seven . . ."

"Six, five, four, three, two, one," the chant went on. A great hush was followed by buzzing, like the sound of a million bees. A giant sphere of plasma sparked and danced around the metal ball. An electrical corona outlined the figure of Dr. Borodine like a three-dimensional Kirlian photograph. A beam of intense green laser light was suddenly emitted from the column, and soared into the night.

The crowd was ecstatic. They joined their leader in turning their faces skyward with a single motion.

A woman standing near Rachel suddenly shouted, pointing at the stars: "There they are! I see them! They're here!"

Those around her reacted excitedly, until a man's flat voice interjected: "That light? That's just a plane," he said.

"But it came out of nowhere!" she insisted.

"No, I saw it make a turn," someone broke in.

"It's blinking! It's sending us signals!"

In a deep voice, another woman asked urgently: "Does anybody here know Morse code?"

"It's no signal, I can see a green light and a red light," a teenager said.

"Flying saucers don't have green and red lights, dummy!"

"Some of 'em do. I read it in a Time-Life book."

A heated debate began. A few said it was undoubtedly a genuine UFO, most now said it wasn't. Gradually the excitement diminished, and when the sound of the engine reached them they finally realized it was only a plane after all.

"Perhaps it's a plane disguised as a UFO. They can do that, ya know," insisted the woman with the sultry voice.

Thomas smiled to himself. Rachel and Peter stood aside, near the two-man camera crew shooting the sky in the background.

"Christ, if they're gonna land, I wish they'd get on with it," said Thomas Archer wearily. "I've got a flight to catch."

There were feverish calls and arguments all around them. Voices in the crowd spoke of seeing saucers, and Borodine could still be heard over the loudspeakers.

"Watch the skiiiiies. Keep watching the skies . . ."

The crowd quieted. Eventually some went off, with the kind of sadness that hangs over the supporters of the losing team after a football game.

"Keep watching . . ." Borodine repeated, late into the night.

But the aliens never came.

* * *

Dawn on Wednesday found the crowd tired, with many sleeping on the rough ground. Borodine was still awake. His eyes were angry and bloodshot. He hadn't stopped talking.

"I'll tell you why they decided not to touch down . . . It's because you people are too damn cheap and selfish!" He yelled at his followers in a raspy, irritated voice.

He was still standing on stage, speaking wearily into the microphone. He was sweating heavily, and his robe was wrinkled. He looked tired and disgusted.

"If you'd just given more to the Cause, *like I told you to,* we could've completed the Landing Strip and given them the welcome they deserved."

In the crowd some people were sullenly preparing to leave. Many were still sleeping, a few were weeping, but dozens were on their feet, solemnly nodding their heads as they listened to their leader.

"You can't expect them to come down without a proper landing strip! Would you invite friends to dinner and only fill half their plate?"

A few of the faithful answered "No!" but the response was weak. It was obvious that the enthusiasm of the night had given way to lassitude. Near the rear of the crowd, lying on his back, Peter was using his jacket as a pillow. Rachel was snuggled up beside him, her arms folded over her blue sweater, sleeping with her head on his chest. Peter stroked her hair and closed his eyes again.

Ellis Denton approached them.

"I sense many skeptics in our midst," Borodine was rambling on. "The negative protons, or negatrons, that emanate from the closed mind of the skeptic, are instantly detected by the Space Brothers. How could they possibly land with a steady barrage of negatrons flying around?"

In the background, Tom Archer and the *Inside Story* crew were packing up their gear, carrying loops of cable and returning the microphones to their cases.

Ellis crouched down, and seemed to relish startling Peter and Rachel:

"Rise and shine, lovebirds!" he nearly screamed into their ears. "Time to go for a drive!"

* * *

An hour later, Ellis's pickup was racing in the desert again. After leaving the BGS encampment they had passed a few isolated service stations, one of which was open, and a dozen large ranches. Denton was driving, and Rachel and Peter sat near him, yawning. Keller even slept for a while, once opening his eyes in time to catch a blurred view of a sign that read: ENTERING NEVADA.

He shook himself awake. Denton had some coffee in a thermos jug, which he passed around.

"So, how was your meeting with the doctor, Ellis?" Peter finally asked.

"I told Borodine everything," the man answered brightly. "He seemed quite surprised, alarmed even, when I explained to him what kind of experiment had been going on. He protested that his aliens' true space hardware could cause no harm."

"Sounds familiar, doesn't it, Rachel?" Peter observed. "Borodine has his mind made up. After all, he's sure he knows what flying saucers should look like."

Rachel looked straight ahead, her mouth set, her face tense.

"I'm fed up with the Brotherhood," she said. "They're just after the money, exploiting people. I should have seen it before. That farce last night, pretending they were expecting the Visitors ... I don't wanna hear another word about Borodine. Just get us to the worksite, Ellis. I'm ready for the real thing."

They kept driving in silence, stopping only once at a tiny roadside café set up in an old trailer. After that the pickup left the asphalt and turned down a barren, rocky dirt trail. It was now mid-afternoon, and long shadows gave the Nevada landscape an eerie appearance. Rachel looked out at the road with eager anticipation. It seemed to run straight into a hellish-looking mountain range, and there was not a living thing in sight.

"One thing I'd like to know," Ellis began. "Who are you two? I mean who are you, really?"

"We've told you who we are," answered Rachel, who felt her anger begin to return. "Why can't you believe us?"

"Well for one thing ... you two don't make sense. Rachel, for instance: You say you hardly know Keller, yet here you are. Why?"

Rachel answered, looking straight ahead, speaking to Ellis as if Keller weren't there next to her.

"Because I happen to like Pete," she said simply. "I like him a lot. And I know him a lot better than he thinks I do."

Peter was caught off-guard by her candid comment.

She looked down as she went on: "He once said he was sorry he hurt me. I could tell apologizing wasn't easy for him ... But somehow I sensed he wouldn't lie to me. To himself maybe, but not to me. Does that answer your question?"

The scientist shook his head and kept driving in silence, until they reached a roadside diner that seemed to have come straight out of the fifties.

* * *

They had driven a few more hours, and sunset was approaching when they saw a Jeep in the distance ahead of them, parked by the side of the highway. Ellis tensed at the wheel, and his eyes briefly looked up to his rifles. Peter put his hand on his arm to reassure him: "Relax. I think I know who they are."

As they were about to pass the vehicle, a man and a woman jumped out and moved to the middle of the road, motioning for the pickup to stop.

"I know them!" said Peter with excitement. "It's about time they showed up."

"Your friends the Rogues?" asked Denton skeptically. "Tell me the truth, now. Don't play any games with me, Keller."

"That's right. The guy is Colonel Trent; I already told you about him. The woman's name is Junko."

Reluctantly, Denton brought the truck to a halt. He shook his head.

"I don't like it. The five of us stick out like a sore thumb in this desert," he said with a snort. "Quite a discreet little expedition you've got going."

Ignoring the remark, Peter greeted the pair and introduced Rachel. Ellis came out of the truck long enough to shake hands with Trent and Junko.

"At least you can see I wasn't lying to you," said Peter with gushing frankness. "They're real. They'll help us unravel this whole business."

"You'll have some unraveling to do, all right," Ellis told Colonel Trent. "From what Keller tells me, I could use an explanation or two about those games Alintel's playing. Assuming Alintel's real."

Junko cut him off. "Let's not waste any time in idle talk," she said. "Who knows if they're not looking for us right now?"

Trent scanned the sky nervously.

"How far is that SCIF of yours?" he asked Ellis.

"Another couple of hours, maybe a bit more."

"Then we can reach it and set up our camp before nightfall. We'll follow you in the Jeep. We'll have plenty of time to tell ghost stories around the fire, Mr. Denton, and to reminisce about UFOs."

* * *

The pickup was now parked in a desolate ravine at the mouth of a box canyon. Ellis and Peter were crouched down, examining the seemingly undisturbed ground; Rachel wandered off in the background. Colonel Trent had parked his Jeep behind a rocky outcropping.

Things were not going well.

"This was the staging area. I'm sure this is the spot: there were buildings right here, cars, trucks ... they must've vacuumed the ground," Ellis said, crestfallen.

He looked at Peter, who was obviously skeptical.

"I'm not lying to you, Keller. It was all here, four years ago. As big as the Manhattan Project. Top scientists, security personnel ..."

"And all of you thought you were working on Star Wars?" Colonel Trent enquired.

"Yes, but it didn't add up. We were doing strange tests on gravitational fields and energy displacement. We were totally compartmentalized, but I think the other scientists were as puzzled as I was. Every day, huge trucks and buses with blacked-out windows would come and go, always to the south. Now, I'd seen maps, and I knew there was nothing to the south . . . not above-ground anyway."

Ellis dropped down and put his ear to the earth. He rose again, shaking his head.

"Sometimes . . . You'd be able to hear the rumble of machinery . . . coming from beneath the surface."

"These trucks—did you ever find out where they were going?" Keller asked.

Ellis looked around as if the memory were putting him on edge.

"Yeah, I found out—and I've spent the rest of my life wishing I hadn't."

Peter pushed him: "The base . . . You saw the underground base, right? Where is it, Ellis?"

Ellis reluctantly walked forward until he was outside the mouth of the canyon; he pointed across the rolling sand to an oddly shaped hill on the far horizon.

"I might as well tell you, it's partially artificial. We called it Pyramid Mountain. The facility you're looking for is directly below it . . ."

He suddenly tensed: "Did you hear something?"

Ellis had his head cocked; he turned in a circle, scanning the desert suspiciously, then suddenly panicked, looking frantically at something in the sky. Peter spun around to see a low-flying black machine coming straight at them. It was ominously, eerily quiet, a cross between an airplane and a jet helicopter. It touched down a few scant yards beyond their position.

Ellis began to back away, frozen in fear. He watched the doors of the craft open and its two occupants climb out. They were not in uniform. One of them sported a dark beard and looked like a graduate student on a field trip. The other one looked older. He carried a gun. In one smooth motion, he aimed it as Denton's forehead.

"You bastard . . . you set me up, didn't you?" Ellis had time to scream at Keller, trying to back away, before a shot rang out. He fell to the ground, his head a mess of red, red everywhere, and Rachel cried out in horror as the blood splattered her. Peter jumped forward, mad with rage and terror, his heart pumping adrenaline to every part of his body. But the assassin was already moving back towards the aircraft, while the bearded fellow held the group in check with a pistol.

Undaunted, Trent returned fire and missed while Junko raced madly towards the plane, rushing their assailants. Everything happened very fast after that: A muffled roar, quickly followed by another; red stains spreading across Junko's blouse. She staggered, landed on her knees and fell forward, staring up at the bearded man with a look of puzzlement frozen on her face. Trent ran towards her and stumbled as a bullet flew right past his head. Another bullet hit the Jeep's engine. Liquid spurted from a broken hose, or a punctured radiator.

The pilot rushed towards Junko's body, dragging her away into the plane while the bearded man kept Trent, Rachel and Keller in check. Seconds later the machine had taken off, Trent still firing at its receding shape. The only traces left of the struggle in the darkening desert were a pool of Junko's blood and the body of Ellis Denton, sprawled in the sand. As for the craft that had attacked them, it had hardly made an indentation.

Trent surveyed the scene, reluctantly putting away his weapon.

"We'd better hurry," he told Keller, who was sitting on a boulder, trying to catch his breath. Rachel had turned away from them, and she was on her knees on the hard ground. She was violently sick for several minutes. "Let's bury the poor fellow," Trent said.

"Where are you going to find a shovel?" Keller asked in anger, almost barking.

"We haven't got a shovel. Just help me pile up these rocks over his body, at least, so the birds don't get to him."

They built a sort of mound, as primitive peoples did when they buried their dead, and they topped it up with smaller stones, anxiously watching the sky.

"Let's get out of here," said Trent at last, wiping the sweat off his forehead. "These guys'll be back, and this time they won't leave any of us behind."

PART THREE

EVIDENCE

It is not starvation, not microbes, not cancer, but man himself who is mankind's greatest danger, because he has no adequate protection against psychic epidemics, which are infinitely more devastating in their effect than the greatest natural catastrophes.

Carl Gustav Jung

CHAPTER SEVENTEEN

The silence had settled upon them like an invisible screen, separating their little group from the world where normal people lived, breathed and loved. They didn't need to put the thought into words. The temporary absence of any visible threat only made it worse. The obvious fact was that they must die, and that no one would ever know why they had died.

Peter shivered and tried to dispel the feeling of utter loss that crept over him as he looked at Ellis's pitiful truck, their only chance to move through the desert—but towards what elusive goal?

"You shouldn't have brought the girl," Colonel Trent said as he unloaded two large boxes from the back of his disabled Jeep. "A serious mistake."

"Why is that?" Keller asked, ready for a confrontation, ready for anything that would take his mind away from the thought of death.

Around them the desert was taking on sunset colors, the sandstone turning ocher and purple. Fine elongated clouds of orange and white stood out in the darkening sky like streamers.

"Complicates the equation. It's a question of trust."

"Ha!" Peter gave a bitter laugh. "And trust is so important to you, isn't it, Trent? After we nearly got blown to pieces ... after what just happened ..."

Trent didn't bother to answer. He carried the two cases into the back of the pickup. Rachel was waiting for them, sitting in silence behind the wheel, breathing softly, trying to recover from the shock, the fight, the spilled blood, the smell of death.

"What's in there?" Keller asked, pointing at the boxes.

"A few weapons. Just insurance . . . I hope we won't need to use any of it."

He motioned for Peter to get in, and for Rachel to drive, the pickup pitching and rolling as it raced toward the pass that led out of the desolate desert valley.

"The Rogues are finished," Trent said somberly. "All the others are dead."

"Everyone but you?"

Trent made a vague gesture, as if he were about to say something, but before he could answer Keller spotted a new threat in the sky.

"Jesus, what the hell are those?" he yelled in alarm, motioning urgently for Trent to look. The man hunched and whistled softly.

"UAVs," he replied without hesitation, "Unmanned Aerial Vehicles. And they're armed. Your basic killing machines."

The two devices were rapidly closing in, coming from the direction of Pyramid Mountain. They skimmed above the desert, low to the ground, zeroing in on the fleeing pickup truck—bizarre, pilotless platforms.

Keller had once done a documentary on model airplanes that included a demonstration of a robot craft, a drone designed for aerial photography. But these models were far more deadly. The one on the left had a body made of twin spheres, with a central axis and rotary blades, while the other one was a cross between a praying mantis and a mini-helicopter. They clearly meant business.

The double sphere must have carried some sort of ranging device, because it hit the hood of the truck with a laser beam and kept a bright spot of red light positioned there. Equipped with deadly machine guns, the other device opened fire as it drew near.

Right away, bullets sprayed the side of the truck, putting a hole in the hood but fortunately missing the vital parts. Rachel never panicked, concentrating on her driving and swerving the truck in a desperate attempt to foul up the aim of the attackers. Trent swiveled his head and got a good look at the objects as they screamed past.

"Drones!" He screamed over the roar of the engine. "Fucking

third-generation drones! They'll cut us to ribbons if we don't take cover."

The UAVs had shot past the pickup. Rachel and Peter could only watch helplessly as the machines made sweeping turns in the sky ahead, in preparation for another run on the truck, again locking their deadly aim on the swerving target.

Rachel kept the pedal to the floor. The devices closed in with ease.

Her face glowed in the reflected light, and Peter felt his heart sink. One truth had emerged from the ordeal: He realized how much she meant to him, and that he might never have a chance to tell her.

"Hang on!" she warned. And just when she judged the drones were about to open fire, she yanked the wheel sharply to the right. The truck nearly overturned as it bounced over a sand dune. This time bullets shredded the tires and blasted the hood. The UAVs streaked by. The pickup lurched to a painful stop.

Hearts pounding, the three fugitives jumped out and made a desperate run for the relative shelter of a nearby rock formation while the killing machines arched over in a dizzying hammerhead turn. Peter took Rachel by the hand and pulled her along, sprinting down a faint dirt trail.

Another thirty yards and they looked behind them, expecting the worst, ready to see the drones coming back for the attack; but the two devices must have changed their tactics because they had retreated and were mere dots far away. Their main objective had likely been the truck, whose metal mass and distinctive heat made it an easy target for the ranging sensors. Peter noticed something else—first an odd glow on the horizon, then a weird superstructure of disks and masts rising in the shimmering heat.

The glow resolved itself into a bank of powerful headlights, and the enormous bulk of the Brotherhood traveling coach seemed to emerge in all its glory from the desert floor. Microwave dishes spun on its roof and a giant silver painting of a fat UFO adorned its right side.

A few minutes later a familiar white-haired head, surmounted by a sea-captain's hat emblazoned with a gold BGS emblem, leaned out the window and stared down at them.

"Well, don't just stand there!" Doctor Borodine ordered with a scowl.

* * *

Peter shivered when he saw that the drones were approaching again. The pickup was out of commission, and there was no time to argue. Carrying his precious camera, he quickly followed Rachel into the coach. As soon as Trent had transferred his boxes from the pickup to the luggage rack on top of the bus, Borodine comically, absurdly took the trouble to switch the destination sign above the windshield, from EARTH PLANE to GALACTIC CENTER to ASTRAL REALM. Then he floored it, knocking his passengers off their feet.

The inside of the coach was an impressive cross between a NASA control room and a plush rock-and-roll touring vehicle. Small models of planets and moons were hanging from the ceiling, creating a lurid science-fiction decor. The vehicle smelled of expensive leather and Old Spice, Borodine's favorite fragrance.

"I never thought I'd be this glad to see you," Keller said, anxiously watching the approaching UAVs through the windshield.

"Well, I'm not glad to see you!" Borodine exploded. "Look what you've dragged me into now!"

The drones were back, flying overhead, criss-crossing the area. Bullets riddled the roof and walls of the bus. They took out one of the headlights, but none got inside.

Borodine cackled with pride, to Peter's surprise.

"This whole wagon's made of kevlar, completely bullet-proof!" he boasted. "The Pope's got one just like it, over in Rome."

Trent didn't seem to share the prophet's enthusiasm. He moved over to a side window and watched anxiously as the probes banked for another strafing run. His face was drawn and frightened, the skin tightly stretched over the cheekbones.

"Helmut, you fool, your tires aren't made of kevlar! We gotta take these damned drones out before we're completely immobilized again."

An indignant Borodine heard the remark and snapped his head back in Trent's direction, but he didn't reply. Sulking, he bent down over the steering wheel and stared straight ahead.

"Slow down a little, Helmut, we've got to get up on the roof," Trent went on. He turned to Keller with a sense of quiet urgency.

"Come on, Pete! I don't want to be a sitting duck!"

They climbed up a ladder, pushed open the hatch and clambered onto the top of the swerving bus, holding onto every possible handle. Trent hurriedly opened the first of his two cases, pulling out a couple of large tubes.

"What are those?" Peter asked.

"Stingers. Shoulder weapons. Here, grab this one!"

He pulled out one of the compact missile launchers and loaded it.

"Watch what I do."

He slid it over to Peter, then hoisted the second launcher onto his own shoulder. Locking his knees against the roof rack, he released the safety latch and took careful aim at the first of the approaching drones.

It opened fire.

Trent remained cool as bullets whizzed past him. He fired back, and the heat-seeking missile streaked across the sky in seconds, hitting the drone dead-on. The device disintegrated, its rotary blade pinwheeling to the ground. The beams from the laser scanned the sky wildly like fireworks, then died.

Peter was impressed and a bit relieved: They seemed to have a thin chance of survival after all. He desperately wanted to hold on to it. But he had no time to offer congratulations: The other UAV was already bearing down on them. Keeping his balance as best he could, Keller quickly lined up the unfamiliar device and fired at the descending drone, but the missile spiraled off-target, missing the approaching machine by thirty yards.

"Not bad. For a beginner," Trent said, his voice a dry croak belying his attempt at humor.

He reloaded his weapon while the flying grasshopper, now deprived of its ranging companion, sprayed the ground with another cascade of bullets, none of which hit the bus.

Trent expertly elevated the missile launcher and fired his rocket at the deadly device, clipping the bottom of the drone and sending it careening off-course. It crashed into the rocks to the right of the racing bus, and exploded in a ball of searing flame.

Peter took a deep breath, shaken by the nightmarish scene. A thousand questions flashed through his mind, questions for which there was no time, and perhaps no answer. Trent, his eyes narrowed as he surveyed the sky, returned the launcher to its case, motioning for Keller to do the same.

* * *

Night had fallen and the bus was parked with its lights out. They had found the refrigerator well stocked with soft drinks and an assortment of chocolate bars, which relieved their hunger. The cabinets above the sink were full of packages of trail mix and potato chips, the prophet's perennial diet, which the fugitives were happy to share.

Borodine was sitting in the cab, lost as if in a dream, perhaps trying to catch telepathic transmissions from his elusive psychic guides. Rachel and Trent were standing a few yards away, exhausted. Peter sat down in the sand, his back against a wheel of the coach, a can of Coke in his hand.

Trent unfolded a topographical survey map and pointed across the dark desert plain:

"Pyramid Mountain ... That's where Denton claimed the SCIF was, yes?"

Peter nodded and he took a sip of his drink, thankful for the coolness.

"It won't be easy; there must be ground sensors all across that plain. But I have some more surprises in store for the Alintel boys," Trent said.

He went back inside and dragged his second case from the bus, revealing a hand-held transmitting device and a dozen weird-looking contraptions. They were foot-long, insect-like robots, featuring six legs and a hammer-like tail.

"These little gadgets are smart. They walk fast and they can negotiate tough terrain." He pointed to the tail, adding: "This thumper is designed to get the sensors' attention. Hopefully it will distract them from us."

Trent set the first microbot on the ground and flicked on the transmitter. The ugly little thing elevated itself six inches and began walking at a brisk pace across the desert. To Peter it looked like a deadly, oversized scorpion.

Rachel walked over to them, her face tense.

"Shouldn't we get going before they send more of those killing machines after us?" she asked, shivering.

Peter got to his feet and turned towards Trent:

"Rachel's right; let's move on while we can."

"We've got a little time," the colonel said. "The drones we destroyed were fully automated. No one'll notice their absence until they fail to return to their base. Even then, there may not be an alarm until they do a ground search. We're still far from their perimeter."

"How much time do we have?" asked Keller.

"A couple of hours at most. That's our window of opportunity. We'll start as soon as the robots are safely deployed ahead of us."

"What's the plan?"

"One, we gotta find the base and get into it without tripping on a mine. Two, we find out what scared the shit out of Gorby's boys. Three, and most important, you've got to tell the world, Pete. And Rachel must be able to corroborate your story. Which means you've got to get out in one piece. We're into this up to your knees, we've got no choice . . ."

He looked at the sky, at the expanse of sand and rock ahead of them and went on: "Our best chance of staying alive is for you guys to make it back to New York with solid evidence on tape, and break the story to the whole world."

They turned as they heard a tapping sound from the direction of the bus: Borodine was impatiently knocking on the driver's-side window and pointing at his space watch.

"What about Doctor Borodine?" asked Rachel. "Doesn't his testimony count?"

Trent dismissed the question with a shrug.

"Of course not. He's useless as a witness. Zero credibility. Alintel created him: staged the sightings for him, the lights in the sky, the whole nine yards. They control many cults that way, from the top on down."

"The sheep and the shepherd, eh, Trent?" Keller said, a look of hatred in his eyes.

Trent remained cool, unperturbed, the earlier fear gone from his face.

"Yes, absolutely. You got it. You're learning something."

"I need some better answers, Trent. A lot better. Forget the New-Age junk. You've already gotten one innocent man killed. You're using all these people like so many puppets."

The colonel looked amused.

"The New Age is the religion of the future, in case you haven't noticed. Isn't that why you guys did an entire show on it? The leaders of Alintel will be its high priests. It will work very well, as it did in ancient Egypt." He grinned and added, "We're the new Pharaohs."

"That's horrendous," Rachel said. "You've used this man Borodine for twenty years ... Isn't it time to tell him the truth?"

"What makes you think he'd believe the truth?" asked the colonel with a grim smile. "He was a nobody when we picked him, and now he's graduated to a full-fledged prophet, with a cushy bank account and an entourage of pretty girls who swoon whenever he opens his mouth. You want to try and take that away from him? Good luck, lady."

Borodine blasted the horn and rolled down his window.

"I don't have all night! Let's move on. The Visitors are waiting!"

Keller shot Trent a pointed look and spoke with his voice lowered.

"What's he doing here anyway?" he asked. "Why did he follow us?"

Trent shrugged: "I had nothing to do with it. Your friend Denton told him all about the base and got him all worked up. So, when he called out to the saucers at the Four Corners and they failed to show up, Borodine must have concluded that his Space Brothers decided to come down at Pyramid Mountain instead of landing at the foot of his crazy Integratron. The rest is history."

Peter cringed as Borodine blasted the horn again.

* * *

The Alintel team had turned down the lights in the Vault, leaving only the large teleconferencing screen and the computer displays activated, so that the features of everyone in the room were tinted with a sickly, greenish glow. General Bushnell was there, along with

Beverly Bernard and Joe Wilkinson, watching the face of the man displayed on the main screen. Greg Parker and the man in black were sitting at another console, monitoring a digital transmission in the form of complicated lines and graphics.

The man on the screen was addressing them from an unknown location. Most Alintel agents only knew him by his code name of Vulcan, a choice Greg found highly appropriate. Vulcan was not only the Roman god of the underground and of industry, but an important planet was once named after him. Its coordinates had often been calculated by nineteenth-century astronomers, yet it was never found in the solar system.

As a planetoid, Vulcan was an unidentified object in the sky. As a major player in world history, the man who now used that name would also remain unidentified. Yet the destiny of the world would change, because of what he was about to tell them.

"I am aware that you are working on many fronts as we approach the date of activation," said the face on the screen, "but that is all the more reason to resist being distracted by small details. I want you to remain focused on our main task. Let me restate our goal, in the context of a cynical appraisal of the lessons of history."

It was a fine face, lined with experience and sorrow, yet alive with the kind of inspiration one would have expected to find in a much younger man. He was sixty, sixty-five perhaps, with a crown of white hair.

"World War I was a clash between large crowds of farmers led by aristocrats. The determining factor was the simple will to gain victory at any cost. World War II was very different, a technical confrontation among large industrial powers. The determining factor was the sheer volume of hardware and the mastery of the science of mass destruction.

"Then the world began to change. Progress was slow at first. We entered a new era, dominated by a combination of mass media and selective small group communications through computer networks. Strategic advantage in future conflicts now resides in the ability of the elite to manipulate mass beliefs."

"But the elite itself has drastically changed in the process," pointed out Bushnell.

"You're right. The new elite has no need to be visible any more," Vulcan said. "In these days of internetworking it can achieve an enormous span of control without revealing itself. It can call the shots from vaults like these, hidden under nondescript office buildings. And the main issue isn't money at all. It is a spiritual one. Look at Catholicism in Poland, Moslem movements in the former Soviet Union, the Islamic revival in North Africa. Against this backdrop the expectation of extraterrestrial visitors is a factor of tremendous strategic significance, all the more explosive since we have managed to keep it repressed for so long."

A beeping sound came from the secondary computer, interrupting the conversation. Greg Parker pushed back his chair and swiveled towards the rest of the group.

"General, we are receiving a new transmission from Pyramid Base," he announced, adjusting his display to highlight a series of graphs. Do you want me to bridge the channels?"

* * *

In the desert plain, the big coach with the ludicrous paintings was moving slowly through the night, rolling without headlights over the perilous ground, following the dozen automated scorpions that were carefully picking their way across the sand dunes. Any stray prospector who might have witnessed the scene could have been excused for fleeing in sheer terror, or for considering a permanent switch to a different brand of whiskey.

When one of the crawlers veered to the left of the bus, perhaps forty yards away, it triggered an explosion that dumped dust and debris all over the vehicle.

"That was a good one," Trent said calmly.

"What you mean is, we nearly blew up on a mine," Peter corrected him, breathing hard.

Trent grunted, "That's a good sign. It means we must be on the right track."

The dark mass of Pyramid Mountain loomed ever larger in the windshield as Borodine now steered with greater attention, zigzagging behind the little automata under Trent's careful guidance. Everyone was tense and expectant except Borodine, who was no longer sulking. He even seemed to have regained his old blustery self, occasionally barking out orders from behind the wheel.

"Keep your eyes starboard!" he ordered. "I sense evil forces at work around here, forces trying to keep me from my destiny. But this time I will not be denied!"

"Better keep your eyes on the landscape, buddy," said the colonel. "The monotony is deceptive."

"Ah, but knowest thou the Ordinances of Heaven?" asked the prophet, quoting the Book of Job.

The biblical reference was lost on Colonel Trent, who kept scanning the darkness ahead in frowning worry. But it was Rachel who first noticed a bunker-like construction whose color and texture blended well into the hillside. This low-profile structure came into better view as the bullet-scarred bus quietly approached the base of the mountain, slowing to a crawl when the terrain became too rough for its small-diameter wheels.

The robot creatures fanned out towards the structure, and halted one after another. Now the building was dead ahead, on the left side of the mountain. The bus rolled to a stop near a rock outcropping.

Slowly, cautiously, the four occupants got out of the coach and began the climb toward the structure, all the while keeping an eye out for any sign of movement. Behind them the microbots had folded their legs and seemed to have gone to sleep. Walking next to Rachel, Keller watched the brilliant stars above them and breathed the cooling desert air. After what they had endured, the scene felt eerily peaceful.

The group moved forward with renewed hope, expecting to see something promising—but they didn't have far to go to realize their mistake. As soon as they reached the fence, Trent's flashlight illuminated the rusting edifice and they saw multiple signs, their paint peeling away in the breeze, that proclaimed the facility was an underground nuclear dumpsite, a depository for highly hazardous radioactive waste. Purple and yellow notices bearing the ominous triangular nuclear hazard symbol told trespassers in no uncertain terms that they were entering an area of deadly radiation. The building itself looked time-worn; aged, crumpled barbed-wire surrounded much of its perimeter.

Peter and Rachel looked to Trent anxiously, hoping he would take charge, but the colonel was frowning, struggling with uncertainty.

"Could this really be the place they brought Denton to, when he worked on SDI?" he asked. "I didn't expect anything like this. The building looks much too old."

"I do think it's the right place," said Rachel.

"You *think* so? Is that the best you can do? Didn't you guys get an accurate description from Denton?"

"I told you the whole story," Peter broke in. "They brought him to the base in a blacked-out bus! His calculations indicated that the base had to be somewhere under Pyramid Mountain. But . . . I don't know what to make of this," he added, pointing at a nuclear warning sign. "Aren't we walking into enough radiation to cook us ten times over?"

"I don't think we have much of a choice, do we?" Trent said, with an engaging smile that belied the dangers ahead.

* * *

"They seem to be having trouble with the Fastwalker again," Greg Parker reported. "They're backing up all the computers at Pyramid Base."

In the darkened Alintel room, its metal walls reflecting the flickering digital screens, the group had reached a decision point. Bushnell and the man on screen they called Vulcan were debating whether or not to proceed with the next phase of their operation. And now some critical data seemed to be missing.

The prize was control, Greg reminded himself. Not just control of America, but long-term control of the planet. And Alintel had the Fastwalker. So why continue the charade, why the elaborate games? Wouldn't it be simpler to tell the people the truth, right now?

But what was the truth?

Parker thought back about the aborigine they had picked up in New Guinea in one of their cross-cultural experiments. If they found him again, how would they ever be able to tell him the truth, even with the best intentions? Should they drag him back to Hollywood and explain to him, in his own language, how he had been fooled by a very long loop of photographs with little holes along the sides? That kind of "truth" would merely drive the poor fellow mad. It would not advance him intellectually or spiritually.

The public was about to be placed in a similar position.

Parker's eyes went back to the display on the workstation in front of him. The group was waiting for his next words.

"If I'm reading this right, the Fastwalker has stopped generating detectable magnetic effects. It has become inert, except for surface phenomena which resemble very fast cell convection."

"What you say doesn't make any sense," said Vulcan sharply. "Surely you realize that. We're dealing with a nuts-and-bolts craft, weighted at over a hundred tons."

Parker checked the screen again, verified the data and went on respectfully, hoping the older man wouldn't notice the excitement in his voice: "The physics group reports the object has gone to a dark phase, sir. Its effective temperature has reached 15 degrees centigrade. They have abandoned the hypothesis that it might represent a lifeform. It is definitely manufactured, just as you stated, but it is still resisting every mechanical test they've subjected it to."

"Tell them to rerun the Doppler tests," ordered General Bushnell. "We'll stay here until we get the results."

Parker turned back to his computer and bent over his keyboard to send the Nevada technical team its new set of instructions.

CHAPTER EIGHTEEN

The weather had eaten through the metal and turned the timbers to a rotting mass. Colonel Trent snapped away rusty chains from the door frame and yanked it open. They found nothing behind it but a steel staircase leading underground.

Peter breathed the stale air, aware of the silence, the coolness. The stairs went down several levels before stopping at a thick iron door with another warning sign:

> DANGER! AUTHORIZED PERSONNEL ONLY
> NBC RULES BEYOND THIS POINT

"What's NBC?" Peter enquired. "I don't suppose they're talking about television."

Trent laughed. " 'Nuclear, Biological and Chemical Warfare' rules is what they mean," he replied. "Proceed at your own risk, if you don't have the masks and the special suits. Not to mention the immunizations."

"No wonder they don't need to post any guards upstairs . . ." mused Peter. "They must be saving tons of money."

A Geiger counter was hanging on the wall. Rachel took it down. They heard the telltale crackle indicating the presence of heavy background radiation, whenever she ran it by the closed door.

"If we stay here much longer . . . we don't stand much of a chance," she said.

They hesitated for another minute, arguing among themselves, soberly considering the lethal possibilities. It was Borodine who broke the stalemate. He suddenly seemed inspired to take charge, went to the door, and declared pompously: "We have nothing to fear, my friends; the Higher Forces are still protecting me. Have they ever abandoned me before?"

Evidently he had already forgotten his dismal failure at the Four Corners. Rachel, Peter and Trent exchanged ironic looks. But he turned the pressure lock without hesitation, and the door swiveled open with a hiss. He stepped through; the others looked at one another, took deep breaths and followed suit.

The four intruders went down a corridor that opened into the side of a large tunnel, dimly lit and devoid of identifying markers. It sloped steadily downward—anyone following it would be taken deeper and deeper underground, with no end in sight. A faint smell of wet earth, of old decaying things, wafted up towards them. Rachel noticed some monorail tracks that ran down the center of the tunnel.

"What purpose does this serve?" she asked, worried.

"That track could have been used to cart away radioactive waste," said Trent coldly. "Or worse."

Keller looked down the long tunnel, which led to God-knew-where. He saw no sign, no opening. All was silence around them.

He took another deep breath.

"We can't turn around now," he finally said. "If the base exists, as Ellis told us it did, it must be somewhere to the right, below the mountain itself."

They began the cautious march down the mouth of the tunnel.

They trekked on, staying close together. Down, always down, moving further away from the surface of the Earth. The air became hot and damp, the smell of decay more intense. They had to fight the lonely, creeping feeling that hundreds of feet of earth and rock were pressing down on them. When they reached a difficult crossroads, a place where the tunnel branched off in two directions, the group halted again in disarray.

"The left tunnel doesn't look as steep," Peter pointed out. "Might be easier to travel."

But Borodine disagreed again.

"Not according to my psychic impressions," he protested, motioning energetically to the right. "It's got to be this way. The Visitors are near us. I can always feel it when I'm close to the EBEs."

"What's an EBE?" Rachel asked under her breath, awed in spite of herself by the man's assurance.

"Extraterrestrial Biological Entity," whispered Peter. "An alien."

He felt her hand shudder in his grasp, the terror of her abduction rushing back to her mind.

Borodine had already started down to the right without waiting for them. Peter started to argue. Trent cut him off.

"These fucking tunnels could lead anywhere. His hunch is as good as any," he said impatiently.

Keller nodded reluctantly, squeezed Rachel's hand in weak reassurance and continued down the right path, several steps behind Borodine, who was quickening his pace.

"Trent was right, I never should've brought you here," he told Rachel.

She quieted him, putting a reassuring finger to his lips.

"Shhh . . . I came because I wanted to. Because I wanted to be with you . . . because you're trying to stop these people, all this evil."

Peter shrugged wearily. "I'm beginning to doubt that we can make a difference."

"At least we can try. When I was a little girl, my older sister Bonnie and I used to go fishing. There was a creek behind our house. But we never caught anything except once, when we pulled out a small catfish with ugly whiskers. He looked at us with big round eyes, as if we were incredibly crazy, or incredibly cruel. We never went fishing after that."

"And now somebody, or something, is fishing *us*..."

"That's why we need to find out who they are. And stop them."

"If we ever get out of this nightmare alive, promise me . . ."

"Listen . . ."

She stopped in mid-sentence and froze. There was a new sound in the stale air of the tunnel, a far-off sound, faint at first but growing steadily louder, disturbing, deep and rumbling. Borodine heard it too: Something big was rapidly approaching, coming down the tracks behind them.

* * *

They began to run, instinctively moving away from the sound as fast as they could. A huge machine was visible now, approaching at a rapid rate. It was moving at high speed, following the track, bearing down on them.

Keller pulled Rachel to him. Together they flattened themselves against the wall, not knowing whether they would live or die in the next moments. Quickly, secretly, her lips brushed against his. The machine blasted by, followed by a compression effect that made their ears buzz, like a fast ride in an elevator. They caught a fleeting look: The monster was as big as two locomotives, with sleek futuristic lines, black-tinted window panels—a cross between a very large bus and a railcar capable of heavy transport.

The tunnel train never slowed, rumbling on until it disappeared from sight. Their companions were safe, having found their own shelter to escape being crushed.

"Do you think they saw us?" Rachel asked.

"I doubt it. I'd bet it's automated," Trent replied. "No humans on board. Nothing but damn robots around here."

Borodine had resumed his march, moving ahead without waiting for the others or looking back.

"It's not far now," he called. "Hurry. The Space Beings are expecting us."

"The blind leading the blind," Keller joked.

Eventually they caught up with Borodine. His legs were weary, but his face was flush with excitement, and he pointed to a glow ahead, a busy combination of flickering lights and colored flashes in the luminous haze that spilled around them.

They were reaching the mouth of the tunnel at last, where the rock suddenly widened. Moving closer to the opening, they heard new sounds—the hum of machinery, and a confused mix of electronically amplified voices. There was a new smell too, a mixture of ozone and heavy oil. When they came out, crouched down on a concrete ledge, what they overlooked was another world, a realm of metal and light, of unlimited power and terrifying complexity. They were in the lair of the Fastwalker.

* * *

The computer screen in front of Greg Parker had remained idle for nearly half an hour. The team had been refining its strategy and

Vulcan expanded on his views about what he called "the extraterrestrial expectation."

"Many individuals who have undergone a sudden contact with an apparently superior level of consciousness have had a major subsequent impact on world history. Without going all the way back to Saul of Tarsus on the road to Damascus, or Joan of Arc, who heard voices in Lorraine and was moved to take over the leadership of the French armies, more recent figures have had a powerful influence. Think of Joseph Smith, the founder of Mormonism, now one of the fastest-growing religions. It all started with a contact with his visit by an angel in New York State, an entity of light."

"John D. Rockefeller, too," General Bushnell broke in. "He saw a blinding flash and had a vision. Voices told him his fortune was a divine trust."

"Here again, the consequences in American life are hard to miss," Vulcan responded. "The Rockefeller Foundation was instrumental in creating the United Nations, new scientific and medical technology, and many aspects of the arts. The Rockefellers founded the University of Chicago."

"Nowadays these manifestations from beyond come in the form of channelings," Joe Wilkinson pointed out in a careful, academic tone.

"Or through UFO abductions, of course. It is possible to control the behavior of a significant portion of the population through this mechanism," Bushnell went on. "A stimulus external to the brain. As we have demonstrated."

He was interrupted by Greg Parker, who was staring at a complex new pattern.

"Sir, we have downloaded a new analysis from the base," said the young analyst. "Their message remains inconclusive, unfortunately. They've established that the craft had no plasma thrusters, and the corona-like effects that showed up on the film shot by Captain James Hall must be due to other causes. The apparent changes of shape we noted on film remain unexplained. There's no obvious reason for the object to be polymorphic."

"What about the luminosity profiles?" shot Bushnell.

"They don't match the analysis of the negative."

"Didn't those idiots measure surface reflectivity?" Vulcan asked.

"Sure did. They say it resembles that of a constantly moving film with an approximate optical depth of five microns, rather than a fixed metallic skin as originally hypothesized."

"What does engineering say about all that?"

"They decline to guess at the nature of the membrane. It seems to be agitated by random interface phenomena at the molecular level."

"Shit! What's it doing now, for Christ's sake?" asked Bushnell.

"It generates a very weak magnetic field, but it doesn't emit any light any more. They're getting ready to run the Doppler tests. Should they postpone the demonstration?"

"Hell, no!" said Vulcan. "Didn't I tell you we needed to remain focused? Think of the political side. By this time tomorrow we must be ready to activate the final steps."

He logged off the channel, and the big screen went blank. The people in the room found themselves concentrating, once again, on the displays at their workstations. The only sound was the occasional click of a key, or the soft hiss of a high-speed drive. Regarding the world outside the room they had no direct knowledge any more, and no interest.

CHAPTER NINETEEN

For Rachel, the most striking aspect of the base, the feature that made her dizzy and induced her to doubt her own senses was the sheer complexity of the sounds she heard. A mammoth underground structure was laid out below her, its dark, far wall only faintly outlined behind a jumble of machinery. In selected places were painted strange symbols, like markers, adjacent to English words and phrases, as though something were being stated in an unknown language. But it was the noise that overwhelmed her.

There was a whirlwind of activity directly before them. Dozens of scientists and military men moved about in intense preparation for something huge.

The train must have followed a different branch of the tunnel from the one along which they had walked, because it had reached the floor of the cavern, with a series of whistles and hisses. Workers in blue berets were unloading dozens of items: lead shields, titanium blocks, test tubes filled with various preparations, steel bars, huge mirrors. They carried a large plexiglass tank holding a blue liquid, and even a screeching chimpanzee in a cage.

To their right was the most magnificent sight of all: a grounded saucer a hundred feet in diameter, silvery-gray in color, with an occasional glint of gold. The sound that came from it was a high-pitch melody that followed none of the known laws of music. The craft's surface shimmered under the spotlights, as if it were spinning rapidly. The disk had a superstructure like a cupola on its upper shell, with pulsing colors that resolved into successive levels of blue, red and green.

The UFO was softly humming and, most incredibly, its physical shape seemed to change ever so slightly from minute to minute, defying perspective, inducing a mild sense of vertigo in anyone who watched it.

* * *

They had been staring at the scene for some five or ten minutes when a new vibration was born. It rose in pitch and filled the space around them. Test devices were being powered up. The nighttime crew was assembling. They came from bunkers through large steel doors.

The workers wore full-body suits with helmets and goggles. They held complex tools in their hands, and filed into the black cavern like so many crawling bugs following mysterious rules of their own. Water was streaming from the ceiling, down an entire wall to the left, the byproduct of a cooling system or energy generator, and it gathered with the sound of a cataract into an excavated drain that led to a sewer-like opening.

To the right, the pressurized entrance to a control chamber formed a gaping zone of darkness. Gleaming plasma devices fed by black cables as thick as a man's arm were aimed at the Fastwalker.

The whole place was absurdly grandiose.

On top of the bunker Rachel saw another ledge, with a concrete balcony running thirty feet above the floor. An elevator was suspended from cables in back of this ledge. On its platform rested two black helicopters, ready to be lifted to the surface. She visualized well-camouflaged openings, hidden shafts that must lead to the desert above.

The Fastwalker had changed shape again, and it began to glow with the intensity of a blaze. What they saw now was a rounded mass of light ninety feet wide and forty feet high. Its intensity resolved itself into successive layers of bluish and reddish radiation.

Rachel squeezed Peter's arm, apprehensive and deeply troubled by the sight. Keller spoke in an awed whisper: "Unbelievable ... How could we ever describe this? I suppose that's ..."

Trent completed his thought:

"That one's the real thing, my friend. A genuine Fastwalker." He let the words sink in, then added: "Nobody could ever simulate this. Not even Alintel."

* * *

In the course of his journalistic assignments, Peter Keller had watched countless interviews of UFO witnesses who had tried to describe flashing lights that were not like the lights on an airplane, colors "that we don't have on Earth," and motions that followed no discernible sequence. They were often at a loss for words. He had discounted their testimony as flaky, or simply too imprecise. Now, face-to-face with the Fastwalker, Peter understood what they had been trying to say.

Color was an intrinsic attribute of the object. It condensed and dispersed, and condensed again in odd patterns. Once, it pulsated in such a way that the entire cave suddenly seemed ready to catch on fire. The operators of the big plasma guns tensed behind their weapons. But the dangerous Fastwalker appeared to have spent all its energy. It became dark again, leaving a formation resembling a glowing grid of embers, orange dots that floated across space, slowly swept aside by air currents. A faint melody began and grew, very different from the shrill vibration they had heard earlier.

It took Keller a few minutes to realize that the shape of the object was changing again. Now it was definitely a disk, its surface shiny and smooth. It had resumed its spinning. Yet it had a bumpy structure on its upper surface, and that bump did not move at all. Rachel and Peter, standing side-by-side near their companions, watched the spectacle with fascination.

To her it was a question of revenge. An unknown force had intruded into her life, taken her away against her will. It had forever disrupted the pattern of her consciousness. To track that force down into the ultimate recesses of her soul, into the secret places where it was hiding, was the only way she knew of regaining her freedom and her sanity.

Peter, too, intended to wrench the secret away, and thus to vindicate an elusive image within himself. It had become far more than a professional obligation to find that truth. It was a sacred purpose now to go after the answer and to track it until the very end, no matter how horrible, dangerous or unsettling it might turn out to be. He clutched his camcorder, checked the battery level for the hundredth time, and secured his back against the rock for a steady shot of the Fastwalker.

* * *

Another phase of activity had begun in the cavern. A door opened. A group of formally-dressed men and women were ushered onto the floor, about ten of them. Several of them looked European, but there were Africans among them, Asians as well, and even an Arab wearing a burnoose. They sat down in folding chairs, like well-behaved children, and they were addressed by a deep male voice that seemed to come from the ceiling of the cavern.

"Welcome, ladies and gentlemen," said the voice. "What you are about to witness will be as hard to believe for diplomats like yourselves as it was at first for us, scientists and military men. We recognize it fully. Our new friends, America's new friends,"—the voice paused here to let the words have their full impact—"have given us a powerful energy source, in exchange for certain considerations."

The guests looked around them in awe, and shuffled in their seats. The voice went on.

"You have already been briefed about the extraordinary nature of these objects. You have seen the film of its interception by one of our jets. Now you are actually in its presence. The power in this room is virtually unlimited. We don't want you to take our word for it. You are about to see for yourselves."

Anticipation rippled through the small audience with a visible thrill.

"No wonder the Reds fell to their knees. Jesus, Lord in Heaven . . ." Trent mumbled. Keller had shouldered his video camera, and went on filming.

A team of engineers rolled a large display screen onto the floor. It was loaded with graphs, words and numbers. Loudspeakers gave sharp, precise instructions. A small group of scientists were looking at monitors in an upstairs gallery. Others clustered around spectrographs, interferometers and precision magnetometers. Laser beams were bouncing off the disk's surface. The display read in large green letters: TEMPERATURE 18.53 C.

"We will now measure the Doppler effect," said the voice from the ceiling.

The display on the screen shifted.

DOPPLER EFFECT CALCULATIONS: VELOCITY = 1,801 km/h, it read.

A hush fell over the assembled specialists. Trent pushed Keller's elbow, pointing at the screen in astonishment.

Peter shrugged. "I'm not good at physics," he confessed.

"You don't have to be an Einstein to see how they're bouncing a beam off the surface of that craft to measure the speed with which it's moving. Problem is, that baby's flying away from us at over a thousand miles an hour, right now."

"But it's not moving at all," Keller protested, looking at Trent as if he had gone mad.

"The disk isn't moving, but its surface is. The skin of the Fastwalker seems to be moving away from us very fast, yet it stays at the same place in front of us," whispered the colonel, astonished. "It's utterly absurd . . ."

"Stand by for destructive testing. Twenty seconds," called out the voice.

"You've got to film this," said the colonel. "I've read about it in classified reports. Never saw it. The material produces its own field, under certain conditions. It's supposed to destroy everything around it."

A palpable sense of danger filled the structure. This was unprecedented, mad, out of proportion to anything Keller had ever experienced or imagined.

"Secure all target items," said the voice. "We have initial coalescence. Ten, nine, eight . . ."

Peter's thoughts went away, to the outside world of normalcy and everyday life, the world where mothers drove their kids to piano lessons, where young couples went to the movies or to a basketball game.

"All systems operators, switch your processors to battle-short condition. Prepare to commence test: Five, four, three, two, one . . ."

The whole structure turned blood-red. A violent energy pulse shot out of the saucer. In a chilling display of immense power, the red beam branched out in a hundred threads of lightning, a spiderweb pattern of awesome destruction: The bars, the mirrors, the blocks, the plexiglass tank and the captive chimp vanished instantaneously, as if they had simply been vaporized.

After a minute or so of stunned silence, applause and cheers rang out from many of the participants.

But the demonstration seemed to have the opposite effect on Helmut Borodine, who had been watching the scene in silence from the back of the ledge. Keller, who had forgotten all about the old prophet, was shocked when he saw him jump out, his face red, veins visible on his temples. He made a sudden move out of their hiding place, and bellowed over the chaotic sounds that came from the machinery.

"Shame on you people! The Space Brothers abhor such violence! They have no need for evil weapons of destruction!"

Trent jumped up and barely managed to overpower Borodine, putting an arm around his neck and a hand over his mouth.

"Will you shut up, you old fool?" he hissed, yanking him back to a crouching position on the ledge. "Are you trying to get us all killed?"

Keller looked around in alarm, and was relieved to see that none of the guards had stirred. The noise level in the cave was so high that no one had heard the old man's outburst over the whir of the machinery and the tumult of excited conversations. Besides, a new group of participants were holding everyone's attention. Trembling, at the edge of tears, Rachel pointed towards them as they came into view.

"Oh Jesus . . ." she wailed softly, gesturing towards a far corner of the cavern, clutching Keller's arm, half hiding behind his body.

Peter turned around in time to watch a dozen short humanoids, slender beings with ugly, deformed heads and huge black eyes, emerging from some unseen part of the base. Their flesh was ivory-white, pasty, with grayish folds, hideously wrinkled. Yet there was a sense of majesty about the deliberate character of their gestures.

The aliens moved slowly, purposefully around the floor, in concert with the Alintel staff. They seemed to be working in concert with the scientists in white uniforms who were leading them among the instruments scattered around the Fastwalker.

* * *

Trent looked at the scene, the whole incredible spectacle of alien beings in the company of the scientists, and he whistled under his breath. "God Almighty . . ." he said. "So that's what's going on! No wonder they kept most of us peons away from the real research. I've heard rumors about . . . about a treaty, a classified arrangement. I didn't believe it, but. . . ."

"What are you telling us?" Keller asked, grabbing his camera, focusing it on the colonel's face. "Spit it out! This is no time to play games."

"It's the deal Ellis Denton kept referring to, don't you understand? The horrible truth he was babbling about. Only worse. It's like this: The aliens have given us their technology . . ." Trent was obviously uncomfortable as he chose his words. "And Alintel works alongside them."

Rachel gasped in disbelief. She looked down at the congenial mix of humans and aliens on the floor of the base, then she seemed to be overpowered by outrage, and grabbed Trent's jacket, shaking him in livid hatred.

"You bastard!!"

Keller kept filming.

"Save your anger for Alintel," Trent went on, shaking her off. "I had no part of this. How was I supposed to know this was going on?"

Peter urgently interrupted their confrontation.

"Stop arguing, dammit! Let me record that scene. We'll have to prove we were really here, inside the base, that it wasn't just some fantasy. Or a dream of ours."

"Or a nightmare," Rachel corrected him.

Trent nodded his head in agreement, and stealthily guided Peter to the left, where he would have a better vantage point for the camera shots.

But Rachel noticed something else that froze her blood. In the excitement and turmoil caused by the appearance of the aliens, Helmut Borodine had quietly slipped away from the ledge.

This time he was climbing straight down toward the UFO, moving at a deliberate pace. Rachel couldn't risk calling out to him, or to the others. She jumped and raced silently after Borodine just as he was reaching the ground floor, in full view of the guards in blue berets.

The misguided professor ignored the ivory-gray humanoids as if he didn't see them, and went directly towards the saucer itself. By the time Rachel caught up with him, he was standing squarely in front of the Fastwalker, an impressive sight with his arms held out in a universal gesture of greeting, and tears of adoration in his eyes.

"Here I am, my Brothers!" he shouted with trembling emotion.

"It's been much too long since I've beheld your radiant faces," he went on, screaming joyously, the flesh creased around his eyes. "There is so much we need to discuss!"

Helmut Borodine as the self-appointed television prophet of extraterrestrial contact had been faintly ridiculous. Rachel had laughed at his elaborate robes and childish rituals when he urged his followers to send him donations or to exterminate his skeptical adversaries. But there was nothing left of the old clown now, as he dared to stand up to the awesome force before him. He was as majestic as an ancient Druid greeting the solstice sun over megalithic temples, as inspired as an orchestra conductor directing the closing chords of Beethoven's Ninth Symphony. But something even more striking took Rachel's breath away.

The pulsing shape of the alien craft, rising far above him in fantastic veins of intense color, began responding to Borodine's words.

CHAPTER TWENTY

Rachel Rand was feeling panic rising within her. A loud alarm had sounded as soon as Borodine had been spotted on the cavern floor; the short gray beings and the small crowd of diplomats had been ushered off the working area through two different doors as quickly as the dignity of the circumstances permitted.

Now she was alone with Borodine, pinned in the beams of the projectors, and the old man resisted every attempt she was making to pull him away.

"Team One! Team One! Intruder alert!" she heard. "Secure the Fastwalker perimeter," warned the voice of the controllers, booming out of loudspeakers in the ceiling and reverberating around the angles of the rock.

But Borodine remained undaunted. He pushed her away firmly, without anger, and he moved ever closer to the huge disk. Now he was well within the destruction zone. Blue berets raced across the immense floor, determined to intercept him, weapons at the ready. Would they actually shoot? She wondered. Or would they be afraid that a stray bullet might hit the Fastwalker, triggering uncontrollable forces, immense disasters?

The voice fell from the ceiling again, imperious:

"Sir, you are trespassing on a secure military installation. Stop where you stand now. Place your hands behind your head, whoever you are. Or we will use deadly force."

"Dr. Borodine, please, please come back!" Rachel screamed, pulling on his jacket. Borodine ignored her. He kept moving ahead, tears streaming down his face, his eyes focused onto a point in the distance, well beyond her own range of vision.

"Have they trapped you in there, dear Brothers? I shall set you free, in the name of everything sacred in the Universe, as you have set my mind free ..."

Again the Fastwalker seemed to react to his words, suddenly growing larger as he came near. The test devices stationed around it were knocked over, shattering glass and breaking instruments.

From the relative safety of their hiding place Trent and Keller looked down in horror. They saw the blue berets drop to their knees and take aim at Rachel and at Borodine. The voice yelled an ultimate challenge at the crazy intruder:

"Get away from the object! This is your last chance!"

But Borodine still wouldn't listen. Swept ahead by his emotions, he moved with his arms extended, like a man in a trance. Rachel was frozen, looking from the Fastwalker to Borodine and to the guns that were being trained on her. Now the saucer was mutating, twisting and shifting, changing shape in a mind-blowing series of rapid polyhedral transformations.

"Dr. Borodine ... Please ..." she implored him.

He finally responded, looking back at her with a relaxed, raptured expression.

"It's all right, Rachel." He spoke calmly, with a grand gesture. "You see, they've heard me. Everything is under control."

A faint humming sound arose inside the giant craft, whose colors were evolving across the entire spectrum with stupefying effect. Rachel felt such intense vertigo that she fell to the floor on her hands and knees. A memory grew in her mind, a recollection from the hypnotic session with Borodine during her first visit to the Brotherhood. Under his misguided instructions, she had re-lived her abduction, re-lived the horrible feeling that she was totally under the control of the ugly gray entities. She had recalled their cold, clammy touch, and even their fetid breath. The old prophet, unaware of her emotions, had tried to reassure her, repeating his ludicrous litany: "Everything is under control. You will not be harmed ..."

Now they were caught in eerie waves of green and mauve, emanating from an incredible construct that kept changing its shape, with a hundred guards in uniform, scared out of their wits, pointing submachine guns at her. But Borodine found everything perfectly normal. She was so terrified she wanted to laugh bitterly through her tears.

The Fastwalker expanded again, with a shrieking sound. All the security men took a step backwards. In the upper gallery the scientists seemed transfixed.

"Professor, come back ..." Rachel cried. "Listen to me ..."

"It's all under control," he said again.

The syncopated melody went to a higher pitch. A piece of rock detached itself from the ceiling, fell and shattered a computer on the floor. A few technicians seemed to awaken from their stupor, and ran to safety along the side of the cavern. Pieces of luminous machinery, floating structures of light, appeared in mid-air. A haunting electronic sound filled the whole space. Everyone retreated against the walls, the guards dropping their weapons. Parts of the Fastwalker vanished.

The disk shot out a devastating force beam that enveloped Borodine and Rachel from head to toe. They were momentarily paralyzed, locked inside its brutal multi-colored coils until their bodies were dissipated piece by piece. Within a few horrifying seconds, the two of them disintegrated into nothingness.

Then the Fastwalker shrank back to its former size and condition, leaving Keller devastated by what he'd just seen. Instinctively, he had bolted forward as if he could still save Rachel. Trent pulled him back.

"She's gone, Peter," said the colonel. "Gone. There's nothing more you can do for her."

Around them the shrill sound died, replaced by a soft whistle, a hum, a trembling, plaintive chant, hushed like a faraway song.

CHAPTER TWENTY-ONE

Peter Keller's body was drenched with cold sweat, and his state of mind was close to despair. He had sworn to help Rachel get to the truth about the mystery that confronted her; he had convinced her to accompany him on a heroic search for a key witness, and now, as a direct result of his actions, Ellis had been shot and Rachel . . .

"Let's just make sure she didn't die for nothing," Trent said, his arm encircling Keller's shoulders.

Peter angrily tried to shake him off, tears of grief and rage obscuring his vision, but Trent forced him to listen, arguing between his teeth, pointing to the camcorder.

"You've got the tape, Pete, goddammit! It's the real thing, the proof you've been looking for. But it means nothing, nothing at all, if we don't get you out of here alive! Don't you see that?"

Keller understood the urgency of the message, and tried to compose himself. Trent repeated his admonition. Grim-faced, wiping his forehead with his sleeve, Peter finally nodded agreement.

Inside the cavern, technicians were stirring again, now that the greatest danger seemed to have passed. Equipment was being restarted. Some of the broken devices were picked up, replaced, recalibrated. The guards were assembling around their leader, lowering their weapons.

"They'll spot us sooner or later," Keller remarked somberly. "How the hell do we get away, smartass?"

Trent looked around with the eyes of a man trained for combat. He pointed to the tunnel train in the background. The machine had begun moving again.

"Here's our ticket out of this place! Let's grab it!" he whispered urgently, walking towards the edge, motioning for Peter to jump first.

As soon as the huge automaton reached them, Keller made a quick appraisal of its speed, scrambled down and fell on its roof. Trent watched him take hold and made sure he had landed safely. But when he tried to follow him, the train was already picking up speed, hissing and rumbling on its tracks. He badly misjudged his leap, slipped off, tumbled to the ground, got to his feet again and raced after the machine. But his white shirt gave him away. He had made himself an easy target.

One of the blue berets spotted Trent sprinting along the track. He had just reached the back of the train when the alarm sounded. The guards aimed at the running figure, and shots rang out. From the speeding machine, Keller saw Trent getting hit. He stiffened, spun around and fell away from the tracks, bloodstains spreading across the back of his shirt.

The train continued to speed up, threatening to shake Peter's precarious balance. Near panic, he managed to lock his foot against a ridge between the machine's metal plates. He could do nothing more than look back helplessly at the shape of the fallen man receding in the distance.

* * *

Three times in one day Peter Keller had seen people die, killed by the forces that preserved the elusive secret of the Fastwalker, the secret he now held within his camera. Shaking, he cradled the camcorder in his arms, opened it quickly, pulled out the tape and thrust it into his pocket.

Keller turned around in time to see that the train was now heading back into the long tunnel, gaining speed. There was precious little room between the roof of the tunnel and the top of the train. He flattened himself against the hard, curved surface and hung on for the wild ride.

The train blasted through the tunnel with the stowaway clinging to its roof, bare inches from being decapitated. At a turn in the track it leaned so dangerously that Peter frantically tried to grab something more secure to hold onto, found nothing but mere rivets, and lost his camera. The camcorder bounced against the rail and was crushed under the wheels. But the tape was still with him.

Only once before had he felt such deep, hopeless despair.

Peter had been born in the Midwest and had grown up in Chicago, the son of a print shop manager who was an inept businessman and quickly ruined the family. He had been sent to one boarding school after another thanks to meager subsidies from his grandmother. He was twelve when his parents separated, and on that day he understood at once how hard a world he would find when he grew up, and how deep and unfair its mysteries were. Somehow he had always survived.

Whenever he reached a tough spot, he always remembered that day of utter loneliness, and found a hidden level of granite solidity within the confused layers of his being. He forced himself not to think about Rachel, he made sure the tape was safe in his pocket, and he silently swore that the whole world would see it, if only he could survive, if only he got to the surface of the Earth again.

* * *

The tunnel train had slowed to a stop. Stiff and sore from the ride, Peter rolled off its roof and climbed down the side, careful not to make any noise. He found himself in a large warehouse that smelled of dust, lit by yellow lights that hung from the high girders. He could hear the sound of workmen moving around near the front of the train, sliding open its doors, unloading its contents with forklifts. Quietly he sidled up to a deserted staircase and followed it up, away from the tracks.

It led to a vertical concrete shaft, metal rungs running up the sides. He climbed up, opened a heavy hatch and peered out onto a courtyard a few feet from a busy sidewalk.

There were many levels of light above him, elaborate displays of colored bulbs and pulsating tubes of all shapes. After the darkness of the tunnel, they blinded him, and momentarily kept him disoriented. For a moment he wondered if he had somehow reentered the Fastwalker's cave. A warm wind brushed his face, bringing the familiar smells of the desert. Slowly he realized that he was staring at the twinkling lights of a Las Vegas casino.

* * *

It was a casino, all right, but it took Peter a few minutes to realize that the scene was not quite right. It was Wednesday night, yet the normally festive atmosphere of Vegas had clearly turned to a frenzy that verged on panic. Crowds were pouring out of the big hotels, and there were mountains of suitcases on the sidewalk. At a newspaper machine in front of the Mirage Hotel he caught sight of a large headline in the *Las Vegas Sun*.

MOBILIZATION, it read. Simple enough.

Groups were forming everywhere. It took him ten minutes to find an empty cab.

"What's going on?" he asked the driver, once they were on their way to the airport. He was an old Chinese fellow, wrinkled and frail.

"Haven't heard?" the man asked in a heavily-accented voice. "Everybody go home now. 'Cept for soldiers. Look!"

A policeman had halted traffic at a wide intersection. A convoy of Army trucks with camouflage tarps rolled in front of them, interspersed with Jeeps, followed by armor-carrying flatbeds that rumbled in front of the Excalibur Hotel, in absurd contrast to the casino's colorful approximation of medieval battlements encased in neon, illuminated from below by red and green spotlights.

"They close Mexican border," the driver went on when he saw his passenger's astonishment. "Million refugee go out of Mexico City because riots there. Leaders, all fighting among demselves."

"That's crazy . . ."

"Whole planet crayzee, Mistah. Maybe end of world, like preacher say."

The driver turned up the radio. The tense voice of a fellow who was reading the dispatches came through, scratchy and trembling a bit, but clear enough:

". . . and the Mexican government has retreated to Guadalajara. In other news Ukraine has given the Russians one week to return the Crimea to them," he said. "After that delay, they threaten to start firing nuclear missiles at Moscow and Saint Petersburg. In light of this ultimatum, President Templeton is trying to mediate . . ."

"Prayzeedent, sure," said the cab driver with a snicker. "Nobody listen to Prayzeedent."

Keller thought back to what the man in black had told him in the New York subway. And the old professor in the bookstore. Then he thought of Ellis Denton, and of Rachel. He remembered her eyes, her smile, and the gentle way her fingers touched his hair. He buried his face in his hands.

* * *

"We don't require the full cooperation of the occupants, whoever they are," the face of Vulcan stated on the computer teleconferencing screen. "When we unveil the Fastwalker for the world, the media impact will be sufficient. The imagery is powerful enough to place the planet under our control."

"You mean, under the control of the U.S. government, I suppose," volunteered Greg Parker.

"Not exactly. I mean under the control of Alintel. An important distinction."

"For how long?" asked Parker skeptically. "Two months? One month?"

"Not even that," said Vulcan. "We don't need more than two or three weeks to make the impact irreversible. It will be one of the great turning points in history, a watershed."

"Like the landing in Normandy," added Joe Wilkinson, rubbing his hands.

"Indeed, gentlemen. That is the way strategic deception always works: achievement of overwhelming surprise. Pearl Harbor. Hiroshima. Sputnik. Now this."

Parker wasn't satisfied. "Some people will remain skeptical," he pointed out."

"Sure they will. That's where you come in. You and that puppet Keller. Keller's tape. I want every man, woman and child in the world to watch that tape on prime time."

"Scientists will question the tape. They'll demand access ..."

"Belief can swipe away any reservation they may have, at least long enough to get past the point where it won't make any difference. We'll discredit anyone who questions the tape. We'll paint them as tools of the old cover-up."

"Besides," interjected Bushnell, "don't forget we have our own assets inside the National Academy."

"And you're confident that our small group here can make the whole world believe that it is being visited by extraterrestrial beings?" asked Parker. "That they work with us? And that we've got their technology?"

"Why not?" Vulcan answered coldly. "It's a bold move, but it will work long enough to have irreversible consequences. Remember, we're not novices at this. The right hand of Alintel is concerned with strategic deception. In the last forty years we have conditioned the U.S. public, indeed, the world masses, to expect visitors from outer space. That belief has reached its critical level."

"So much for the right hand," agreed Parker. "What does the left hand do?"

Vulcan answered calmly. "The left hand of Alintel has been busy getting hold of real hardware. And we've finally succeeded."

"But we don't fully understand these objects," pointed out Joe Wilkinson.

"That's true, my young friend, but our primary objective remains the same: to secure the hardware, to establish our own power. The rule of Alintel. Once we are well on our way, there'll be time to take apart that technology."

"What if we're biting off more than we can chew?" said Greg. "Are we certain that our left hand always knows what our right hand is doing?"

"Absolutely. The only step that remains is to produce the Fastwalkers as evidence of the invasion. Control will fall into our hands like a ripe fruit. Starting with President Templeton."

"And if he refuses?" asked Wilkinson.

"Confronted with overwhelming power and surprise, no politician would take the risk of refusing," answered the big face on the digital screen. "And I know Henry. Believe me, your president is, first and foremost, a politician."

* * *

Keller had reached the designated airline gate, among the clanking sounds of the slot machines. He was second in line at the counter, looking dusty and bedraggled from his escape, and his boots had begun to hurt his feet. He grieved from the loss of Rachel, and from the death of Ellis Denton: There were so many more questions he would have loved to ask the man! So many things he had missed during the hasty interview in the speeding truck . . . He even felt a pang of sorrow for the death of Colonel Trent. He patted his pocket, where the precious tape lay. During the wild ride in the tunnel he had dropped his video camera, but the tape was safe.

Warily, he scanned the crowd for any sign of Alintel. Would he be smart enough to recognize their agents, to evade them if they came after him?

Through the darkened windows he could see the runways. Dozens of huge Hercules transport planes were parked near the hangars. Green Army helicopters were buzzing the terminal.

Keller found that he was exceedingly tired.

The woman in front of him moved on. He reached the counter.

"Can you put me on standby for your next flight to JFK?" he asked.

CHAPTER TWENTY-TWO

The chimpanzee shrieked again, rattling the bars of his cage. Through the haze, a woman's hand appeared. It opened the cage and released the monkey, which scurried away across the rolling landscape.

The hand belonged to Rachel Rand, who was wide-eyed and, to her considerable surprise, felt very much alive. She looked around and found herself standing on a hill above a dark blue plain, a gusty wind whipping up her clothes. There was a curved wall too, very far away, cutting off her view. A cloud drifted by, briefly obscuring an area littered with objects she recognized as the target items from the test in the base.

The liquid in the tank had turned orange, and seemed to be boiling. Other debris from previous experiments dotted the landscape, including an old Pontiac lying on its side, being earnestly studied by a vague silhouette that gradually assumed the familiar shape of Helmut Borodine.

"Where the hell are we?" he asked abruptly, stumbling over his shadow, his large frame oddly outlined against the changing colors of the sky.

She felt lightheaded, disconnected from her normal thought process, an effect not unlike severe jet lag. She heard her own voice weirdly echoing across what seemed like a metal floor.

"Don't know—know—know . . ."

"We must be inside the Fastwalker," Borodine said tentatively. She noted that he was losing his arrogant, blustery tone. "But where's the crew?"

"How can the inside be so much bigger than the outside?" she asked, trying to keep her balance on the tilting, undulating ground.

Somehow she stayed on her feet. Purple tumbleweeds, blown by heavy winds, seemed to pass right through them: They were drawn toward a sound like churning water, coming from a point up ahead.

They arrived at a raging stream. It wasn't water but a brilliant rush of energy that was flowing upwards, from a chasm in the ground to a bright yellow cloud at the top of the sky. They stepped around it and walked on until a wall of haze enveloped them, a glowing fog that forced them to shield their eyes with the back of their arms.

Vague forms, tall columns of luminous smoke, now stood above them, vibrating, undulating, *alive*. Rachel thought she could make out eyes of limpid radiance. Multiple whispery voices rang out inside their heads, overlapping each other rapidly but with a consistent, hive-like train of thought.

"Hold-where-you-are . . ."

It was not an order, rather a statement of fact, a gentle reminder.

"Greetings!" mumbled Borodine. "What planet are we on?"

He wasn't certain where to direct his voice, so he kept spinning as he spoke.

Only laughter came back from the void where the light beings were.

"Earth is our home port," the voice chanted. "Home port, point of permanent origin. Omphalos."

Omphalos, Rachel reflected. *The navel of the world.*

"That makes no sense . . ." Borodine protested.

"We are from the multiverse," the thought went on patiently. "You have dimension-crossed, little humans."

It was music rather than a voice, a very complicated melody.

"Listen, I don't know who or what you are," stated Borodine, trying to sound tough, "but I came here to see the Space People."

A new sound filled their heads. Again, the closest human feeling it resembled was pure, happy laughter.

"You must have been deceived. By lies from your own kind. Nothing but tricks. Beyond the human level there are many layers of consciousness around the Earth—but no space people."

"What about . . . my mission . . . what about . . . the extraterrestrials?" Borodine asked tensely, stumbling over the words.

"It will be years, many decades of your time, before humans achieve contact with the consciousness of space."

"Are you saying that I was tricked?"

"Watch. Look . . ."

A vortex was being created before them, like a window in mid-air.

* * *

Gazing through the vaporous haze, Rachel caught her breath when she saw silhouettes appear. Men. Among them was Colonel Trent. The Alintel operative was standing near the staging area in the underground base; the dignitaries and the gray aliens she had seen before were all gone.

Trent was wearing a bloodstained shirt. There must have been some sort of fight after she was taken by the Fastwalker, Rachel thought. But when he removed his shirt no injury was visible on his skin. An aide handed him a clean one, and he put it on with casual assurance. Several of the base technicians gathered before the colonel, listening as he commented on their work. Next to him stood a willowy Asian woman of forty or forty-five. Her blouse, too, was stained with blood, yet she seemed none the worse for it.

Rachel recognized her with deepening astonishment: It was Junko, the Alintel Rogue who had supposedly been killed in the desert. Blown away by the man from the aircraft. Why wasn't she dead?

"Congratulations, ladies and gentlemen. And thank you for your help, Junko. The exercise was eighty-percent effective."

"I'd always wanted to be an actress," Junko commented lightly.

"You should be on Broadway."

Relaxed applause came from the group, releasing the tension. On the other side of the window Rachel watched with fascination.

"Let's go on with the debriefing," Trent proposed.

One of the technicians stepped forward.

"Question, sir: We've been wondering why it was necessary to dispose of Ellis Denton? Was it just to give credibility to the Rogues?"

"Only in part. General Bushnell felt he would become a liability if allowed to get into the base with Keller. He might have seen through the theatrics. We couldn't take that chance. We came to the conclusion there were too many variables at that point."

"How did you know they would get in here unharmed?" asked one of the guards.

"You fellows were shooting blanks."

"But the drones had live ammunition. Your group could have been blown to bits in the desert," the man continued.

"The chances of that were small," Trent answered coldly. "Their sensors work in the infrared, so they aim at masses of warm metal, not human beings. I was fairly confident I could shoot them down before they could do real damage, because we had degraded their ranging system. Worse case, the bus would've been disabled. That was a risk we had to take."

"Another question, sir: What happened to the man and woman who got too close to the saucer? I assume they account for the twenty percent that was in error?"

Trent spread his hands in a helpless gesture.

"They got disintegrated, of course. Dead. As dead as that chimp in the cage," he said with a tone of reproach. "This would never have happened if you fellows had secured the perimeter according to the exact instructions you had been given. This one mistake could've jeopardized the entire operation. Fortunately those two were expendable."

"Who were they?"

Trent smiled:

"The man's real name was Ernest Chubb, but he's become better known as Doctor Helmut Borodine. He's one of the New-Age prophets we've created. The woman doesn't matter."

"How will you explain their disappearance?"

Rachel and Borodine leaned forward, looking through the portal in angry amazement. Trent answered the question without hesitation.

"We'll enhance the legend, as we always do when challenged by unforeseen events. We'll tell Borodine's followers that he was abducted by the saucers. That's a fashionable term these days. Very consistent. Very much in line with the Eschaton. Besides, it's the truth, isn't it?"

"Are you sure his followers will buy it?"

Trent laughed.

"Enough of them will. We'll present it as a contribution to the wonderful spiritual transformation the human race is undergoing. He was taken away by the Visitors on a journey of many light-years to their home planet; that's paradise, as far as the faithful are concerned. We'll never need to explain why he never came back," he concluded with a slight grin. "After all, the same thing happened to Elijah, taken up to heaven by a celestial chariot. And Enoch. And Ezekiel. The Bible is full of that stuff. So is the Koran, by the way, and the Vedas in India. Our friend is in pretty good company, as prophets go."

Borodine's eyes turned grim and incredibly angry, his mind a swirling, churning cauldron. His hopes, his dreams, his faith had just been crushed. He felt like the humble welder from Hoboken once again. But he had gone to the other side, he had reached the multiverse. And they could never take that away from him.

The portal shrank down and disappeared.

* * *

"So there are two different kinds of aliens after all," Rachel reflected, staring at the beings of light who now towered above her like genies in the desert. Or like angels. "But what about the Short Grays, back in the cavern? Where are those from?"

The entities responded evasively, sending their thoughts inside her head.

"We will not intervene—interfere—in their affairs. Such beings are not our concern."

"Can't you be more specific?" Her mind went back to all the religious tales of her childhood, about mystics who were reputed to communicate with higher entities, about the Shining Beings of American Indian traditions. They were never specific either. When they deigned speak to humans they expressed themselves in riddles like those of the Sphinx. They didn't answer pointed questions from puny humans if that didn't serve their inscrutable purposes.

"We are only the Overseers," answered the voice in her head. It sounded like it came from very far away. "We commute among dimensions."

Rachel kept moving, along with Borodine, along the enormous desert inside the Fastwalker, and she found herself thinking of Chip Huntley. It was odd: She hadn't thought of Huntley in years.

* * *

Chip had been her sister's first teen-age romance, when Bonnie was fifteen and Rachel thirteen. He was the best student in the school. He was very good in math, and excelled in physics. Whenever he came to their house he didn't talk about cars or movies or football like the other boys. He only spoke of outer space and travels to other planets, of plasma thrusters and antigravity. And he had such a passion about it, such a way of bringing to life the dry reality of science, that it gave her a fantastic thrill.

Chip Huntley's pet theory had to do with hyperspace. There were more than four dimensions, he kept telling Rachel and Bonnie. Time was not real at all. It was just an illusion, a side effect of being human. Hadn't Einstein himself worked on theories of higher dimensions? So had the Russians, Andrei Sakharov, and even Konstantin Tsiolkowsky many years before Sakharov. She had forgotten most of Chip's ideas, but she did remember that much.

When he became older, Chip quickly drifted into the kind of peculiar world very gifted youngsters inhabit when they run out of teachers they can trust and grownups they can talk to. He started playing more and more complicated video games, some of which he programmed himself on his home computer. He spent endless days and nights devising advanced schemes involving "Dungeons and Dragons." His dates with Bonnie became tense and boring. After a while Rachel didn't see him at her house any more.

The day came when Chip didn't show up in school at all.

The police eventually found his body in the basement of the institution, and the coroner diagnosed heart failure brought about by a combination of drugs that were supposed to make poor Chip Huntley even smarter than he already was, allegedly providing access to the many dimensions of the virtual world he was trying so desperately to construct.

Now Rachel wished she had listened more carefully to Chip's strange theories, because she was trapped inside a vast desert that filled an impossible structure of sound and light: a structure that once, from the outside, had looked simply like an advanced spaceship.

* * *

"What about the craft Alintel captured?" she asked the entities, forcing the thought out of her own mind. "Are we still inside? What are the Fastwalkers?"

"Our device was permitted to fall into the hands of your scientists—deliberately—for education—temporary. It was—an experiment. Unfortunately misunderstood. We will recall it very soon."

"What will happen when it takes off again?" she asked in alarm.

"Dislocation—unavoidable. You saw the destruction, on a small scale."

Rachel imagined what would take place if the full power of the Fastwalker were unleashed. She visualized the underground base blowing up, the instruments destroyed, and rivers of energy emanating from the object, flowing into the mountain in all directions, lifting the very desert off the surface of the Earth over dozens of square miles.

"How do you create such devices?" Borodine enquired.

"It is easy for us. We mold the form and the substance—drawing from what you see as void-emptiness, yet not empty. They are windows."

"But what if Alintel discovers how your propulsion system works, before you can take off again?"

The lightness came back, the laughter echoed in the void.

"This is not a machine in your sense. It is an elevator for consciousness, a reality transformer, a dimensional gateway. No propulsion can be found. None exists. None is necessary. Only light. Our kind of light."

"What happened to us? How did you capture us?"

"We didn't capture you. The process is accidental," replied the voice. "In four-dimensional space there is such a thing as a twist of a twist, a rotation of a rotation."

"I'll take your word for it," Borodine said.

The space around them seemed to change, increasing its density.

Large luminous machines of milky transparency were spinning around in synchrony with the sound. Light filled the air, reflected on Rachel's face, on Borodine's clothes.

"How do you know about us?"

"We have watched. Studied."

"Why the abductions?" asked Rachel.

Her mind filled with sadness. Their sadness.

"As we said, the process is accidental. Human mentation doesn't expand to grasp it. Your constructs are twisted by our field. The brain's imagination tries to cope with it—and it fills the gap with its own images."

"How long has such contact been happening?"

"We have always been close to your plane of existence—for centuries. Monitored . . ."

The signal halted, punctuated by surprise, then it resumed at a faster pace:

"Hazard now—Could be locked out. Window closes—We give warning . . ."

Borodine and Rachel exchanged a look of puzzlement. The voice had assumed a new immediacy. Thoughts came in garbled, among overlaid rushes of intense sound and color.

"We will transport you . . . to home plane . . ."

It was a fluid statement, a chaotic feeling born inside them.

"I think they're saying they're gonna try and send us back," Rachel began, turning to the prophet.

Borodine looked around at the bizarre landscape.

"I'm all for it," he said with a shrug.

His voice was still echoing long after his shape had blurred and vanished across the dimensions of the multiverse.

CHAPTER TWENTY-THREE

Ben Simon's role model was Geraldo Rivera. He had watched Geraldo grow from a tabloid hack to a television sensationalist, and finally to a wealthy, influential syndicated opinion leader. These transformations took place through an amazing array of interviews that ranged from the utterly ridiculous to the lurid and the juicy, to the profoundly terrifying. So he stayed late in his office the next day to study the tape Keller had brought back, the tape of the underground base. He was running it on his best VCR, working hard, making careful notes, alternately starting and stopping the motion.

"Wild . . . Utterly wild," he said, musing aloud. "Think of what Geraldo would do with this material!"

He looked across the vast expanse of his desk to where Thomas Archer, Chad Burke and Holly Sukuda were waiting for his reaction. They didn't seem to understand the situation.

"Okay, Tommy. Tell him we'll run it," Ben Simon decided. "I'll provide a personal introduction to the piece. Don't spare the promotion money. We'll have plugs in all the morning papers, we'll run teasers all afternoon, and we'll bracket the six o'clock news. He can tell his story in voice-over, and we'll use the tape to dramatize the narrative. Should be sensational as hell, and it will do marvels for our ratings."

Tommy nodded in lukewarm agreement, but Chad remained clearly uncomfortable with the producer's decision.

"Underground bases, limitless power, aliens teaming up with Uncle Sam to shape the future . . . Come on, Ben, don't tell me you believe Keller's bullshit . . ." he said, his big hands helplessly gesturing in the air.

His boss just laughed.

"Who knows where he took those shots—a Hollywood set, a warehouse? Who gives a shit? All I know's it's a fuckin' great story, and it'll generate big numbers. End of discussion!"

* * *

Letting his colleagues argue about his precious tape, a somber Pete Keller had remained in his office, where he was watching television with rapt concentration. He was tuned to a live broadcast from the Brotherhood of Galactic Science network where a leaderless Brother Ralph, who sounded rather lost and lonely, was mechanically imploring viewers to send money, money and more money. Peter barely looked up when his partner returned from Ben Simon's office with the good news about the show. Thomas Archer was weary and a little worried about his friend's reputation and condition, but he tried to sound upbeat.

"Green light, Pete. Ben's gonna run it. You and I are co-producing. It'll be broadcast tomorrow night on prime time."

Keller nodded solemnly, without taking his eyes off the screen.

"You don't seem so thrilled."

"Why should I be? There's blood all over this story, Tommy. Rachel Rand and Helmut Borodine are dead, blown apart, dispersed as electronic dust. So are two of the Rogues who bravely tried to expose the cover-up, Trent and Junko. As soon as they find out what we've got, Alintel will try and kill me like they killed the bookstore guy and Ellis Denton. Our only hope is getting the story out. Tomorrow's a long, long way away."

"They won't touch you once we've gone nationwide. They wouldn't dare."

Keller looked up at him gratefully, wanting to believe his friend's line of reasoning. It was the same line he had used with Ellis Denton.

"Come on, Pete, I'll buy you dinner," Tommy said.

* * *

On the television set in Peter's deserted office, the voice of an excited Brother Ralph went on addressing the members of the Brotherhood, whose ranks had actually swelled since reports of their leader's sensational abduction had been disseminated. Even those who had become disenchanted after the fiasco at the Four Corners were willing to give the Brotherhood another chance. And a lot more money.

"Friends, believe me, now is the best time to give," implored Borodine's sidekick. "The memorial statue of our beloved Doctor, situated in the Four Corners area that he loved so dearly, and to which so many important, cherished memories are attached, is only half complete."

An artist's rendering appeared, of a gigantic statue of Helmut Borodine intensely watching the sky, rising up from a beautiful high-desert landscape, dominating the podium.

"It's the perfect way for us to remember the Doctor," Brother Ralph was saying, "and to thank him for teaching us the ways of Space Consciousness. But it takes money to make this dream a reality. So call in NOW and make a generous . . ."

Something hissed just downstage to his right, causing Ralph to stop in mid-sentence, astonished. The air seemed to be churning and exploding. Vivid colors streamed in pinwheel patterns. It seemed that a small psychedelic tornado had invaded the studio.

All at once, there was a crack like a quick burst of thunder. Dr. Helmut Borodine in person materialized on stage, stepped out of the smoke and waved to the studio audience.

The well-trained band never missed a beat, immediately breaking into the Galactic Science anthem.

A ripple of emotion flowed through the room. Borodine's disciples were overcome with joy. Tears streamed down amazed, smiling faces. Brother Ralph himself, weeping like a baby, spoke to Borodine through a stage mike.

"Transmogrification! We have Resurrection, Ladies and Gentlemen! A historic moment! Doctor, you have made the journey to the stars and you have returned to us in record time!"

Seized with emotion, Ralph went over to Borodine and dropped down on his knees like a little child, wrapping his arms around the prophet's legs.

"Let go, Ralph," the leader said with anger in his voice. "Can't you see we've got lots of work to do? Nevada's about to blow up. The people need to be warned. There's no time to lose."

He stepped away from his startled disciple as the camera followed him across the stage.

"Where the hell is my blackboard?" he asked impatiently, gesturing to his assistants.

* * *

The television set in Keller's darkened, deserted office now showed a highly agitated Dr. Borodine, furiously drawing on his blackboard, finishing a complex geometric pattern. On the top half of the board, the words *Spacecraft, Extraterrestrial, Interplanetary,* and most conspicuously *Space Brothers* had all been energetically crossed out.

Borodine was still speaking to his audience in his trademark gruff, scowling voice, but the message had drastically changed.

". . . And now I have been to the other Dimension. I've seen it with my own two eyes. I damn well know we've all been deceived! Everything I've been telling you for years was dead wrong, do you hear? Forget the Space Brothers, the golden landing strip, the crystals. That's all crap!"

Brother Ralph and a group of BGS officials were standing in the wings, watching Borodine, horror-stricken.

"We gotta get him off. He's killing the business!" whispered Ralph.

But their leader was still speaking to the camera with considerable passion, scribbling new slogans on the blackboard.

"Real UFOs are not from space at all, *not space as we know it.* I'm here to tell you, things are not so simple in the Cosmos. The saucers are actually using this parallel dimension I spoke of. Forget the old universe you idiots learned about in school, *we live in a multiverse, d'you hear?* And it's an incredible place, believe me."

Borodine finished drawing his diagram, slammed the chalk down and spoke straight into the camera with fire in his eyes.

"One of their craft fell into the dirty hands of our government a week or so ago. It's held captive, right now, this minute, inside an underground base in Nevada, where a bunch of goons and space cadets are trying to unlock its limitless power. But that is suicide! They once used UFO technology to scare the Commies, but now they're really playing with fire. The power source will get out of hand any day, when the Fastwalker takes off, and it can destroy the base! Not only that, it will take half of Nevada with it! Millions will die ..."

Abruptly, Borodine's face disappeared and the camera cut to Brother Ralph, standing downstage. He spoke to the viewers in his calm, soothing voice.

"Ladies and gentlemen, Doctor Borodine is very tired and certainly not himself."

The voice of Borodine still reached the listeners, but it sounded far-off, un-miked, furious.

"Liar! Traitor! Dupe! Millions will die ..." The old man was yelling in vain.

"Please disregard everything our good doctor has been telling you since his dramatic return to the Earth Plane," uttered Ralph, speeding up his delivery. "He needs to rest after his extraordinary journey. Abduction phenomena have often resulted in confused thinking during the period following the episode. Post-traumatic stress syndrome, it's called. Dr. Borodine will feel much better tomorrow. Now sit back and enjoy a fine film, here on the BGS satellite network. And uh, keep those phones ringing ..."

But Borodine barreled back into the frame. Shocked gasps came from the studio audience when he fought his way to the microphone. He seemed furious, eager to resume his fiery preaching. A middle-aged woman, a tall blonde with short-cropped hair, stepped forward, a glass in her hand, stopping his motion.

"Your power drink, Professor. You'll need it. You must be terribly thirsty. It will restore your strength."

Mechanically, Borodine took the glass, drained it and gave it back to Dr. Beverly Bernard. He smacked his lips and rushed forward. He wrestled the mike away from Ralph.

"I was sent back by the Overseers for a single reason, and that is to warn the world! A secret group may have captured a saucer, but they're powerless to control its nature. They'll never be able to use it!"

Now he sounded incoherent and looked crazy, waving his arms in the air like a windmill.

"Heed my words, that craft isn't what it seems, it will take off again. We're all in danger, big time, and by the way, Brother Ralph, you're fired!"

Sounds of a scuffle were heard as the live feed was cutoff, abruptly replaced by the cheerful music and opening titles of the 1960 Jerry Lewis film, *Visit to a Small Planet.*

* * *

President Templeton and the First Lady returned from the Lincoln Center concert at 11:00 that Thursday night. The black limousine stopped smoothly in front of the White House. A Secret Service man reached forward and opened the door for them while another agent switched on his transponder.

"That modern piece was interesting," Templeton said to an aide as he stepped out. "It reminded me of a movie I saw once . . ."

"Can I have a word with you, Henry?" asked Bill Renslow, stepping forward among another group of watchful guards under the portico.

"Don't tell me the Mexican situation has gotten worse," growled Templeton. "Didn't we secure the border?"

"The border is sealed, Henry. From Tijuana to Corpus Christi. No problem there."

"Is it Ukraine? Are they still talking about going nuclear? The Ambassador was at the concert, and told me . . ."

"Relax. We can bring all that under control. It's something else altogether. I have a videotape to show you. And it can't wait."

Templeton turned to his wife, who was shivering on the steps. Her private secretary had brought her a shawl, which she draped over the shoulders of her light coat. The weather had turned cold, and the first hints of winter were in the air.

"I'll join you in a little bit, dear," he said with solicitude. "One last thing to take care of."

* * *

They were in the secure vault, deep under the White House, next to the Situation Room, just the two of them, old childhood friends, Henry Templeton and Bill Renslow. They had grown up together in Kentucky. They had gone canoeing together, and chased the girls and studied together in college, until Templeton went into law and Renslow took up political science. They had met again, years later, when Governor Templeton ran for the White House and needed a campaign manager.

Now the president was being confronted with awesome evidence that a higher power than his own controlled the planet.

When the tape was over, Renslow reached for the stop button on the VCR and the vault became eerily silent. The president was white, his cheeks bloodless, his face drawn and tense.

"Why wasn't I told of this when they briefed me, right after the inauguration?" he asked between his clenched teeth.

"Evidently the need-to-know is controlled by the Alintel group. I didn't hear about it myself until this afternoon."

"The Agency sent over their Keeper of the Weird . . ." protested the president. "You were there with me. Just two days ago. Dammit, isn't he supposed to be cleared for that stuff? He didn't say a word about this."

Renslow looked at his feet.

"He didn't have access either. They kept CIA in the dark. I'm sorry to tell you, the guy didn't know anything."

Templeton covered his face in his hands and when he lifted his head again and looked up, his old friend thought he had aged ten years.

"What happens now?" asked the president. "Get on with it, Bill, for Christ's sake! Are we supposed to take orders from those deformed gray things?"

"You'll have to make a public statement. I'll prepare it for you."

"And foreign heads of state?"

"Their key envoys have seen the technology, as the tape shows. By now every capital must be aware of the aliens."

"Long-term . . ."

"Who knows what plans Alintel has made? All we have are immediate moves, their instructions for the next month or so."

"Their own version of a new power structure . . ." mumbled Templeton. "Is that it? Is that what those ugly little bastards will want?"

"For now their requests are more simple: As a first step they ask that the nuclear stockpiles be fully dismantled, under Alintel's control. They want our military warning systems to be turned off. And they will put their own institutions in place. They want a meeting of central banks to be set up, to consider a readjustment of the major currencies, to avoid a panic. As far as the atomic ban is concerned, I am told Russia will comply. Ukraine and North Korea have tentatively agreed. Britain hasn't bothered to answer. We may have a problem with France and Israel."

"And with a few segments of our own population in the U.S., no doubt," Templeton speculated.

"Precisely. In a day or so we must be prepared to declare martial law in some limited areas, if needed; and to suspend certain basic rights, at least for a short time," said Renslow, looking closely at his old friend.

The president reacted as if he had just seen a dozen snakes gliding into the secure vault.

"What on Earth do you mean?"

Renslow didn't answer. He let the silence speak for itself. There would be plenty of time to work out the details.

The president looked around the room at the stark, copper-lined walls, the bank of gray phones, the dark screens of the computer consoles. He was no longer the most powerful man on Earth. He had never been the most powerful man on Earth. Others were in control of forces above his awareness, far beyond his reach. And all those years he had been kept in ignorance.

That was a hard lesson to accept.

* * *

When the president had left, Bill Renslow remained alone in the vault. Feeling a bit guilty, he locked the door behind Templeton and walked over to the phones.

He had been tempted to level with his old companion at the crucial moment, when he had seen him, fragile in the cold night, shivering on the steps of the White House with all the responsibilities of the world on his shoulders. He had quickly controlled his emotions. Presidents change every four years, he reminded himself. Alintel had been designed for a long-term goal that measured itself across the span of generations, not merely in terms of specific administrations. That was one of the reasons for their charter, keeping their little secrets to themselves without bothering to bring the CIA into their confidence. The Director of Central Intelligence, too, tended to change every four years, swinging with unpredictable political winds. Unpredictability was the one thing Alintel abhorred above all else.

So Renslow quickly enabled a secure circuit, tapping a sequence he had memorized well. A voice came, crystal-clear, through layer upon layer of ultra-fast processors. The wonders of fiber-optics, thought Renslow. Technology was on an unstoppable course, and he loved every minute of it.

"Bushnell here," came the voice.

"General Bushnell, this is Vulcan. I have briefed the president. I have made a believer out of him."

It was hard to miss the pride, the note of power in that statement.

"Then I take it we can proceed?" asked the old military man, standing at attention.

"The ball is in your camp," Renslow assented, trying to control the excitement in his throat. "I suggest you make the appropriate number of copies of Peter Keller's tape for distribution to Congress, to the foreign media and to the embassies. When the story breaks tomorrow morning, the Air Force brass is going to deny everything at first. Templeton will wait for the television program to reach its full effect. We will orchestrate media demands for an end to the cover-up, and the president will appear to bow to public opinion by making a statement. That will be his finest hour. Who could stop us after that?"

CHAPTER TWENTY-FOUR

There isn't much to do outside on a cold Thursday night when you're a kid in Milton, Pennsylvania. So thirteen-year old Johnny Harder and his friends were just riding their bikes around the neighborhood when it happened.

They had been joking about their teachers' little quirks, they had gossiped about other kids in their class, and they had occasionally broken into a short race. It was Johnny's lucky day. His father had had a raise at work, and in a sudden burst of generosity, had bought him a new ten-speed bike, all green and gleaming. The boy had convinced his friends to join him in a test ride to the nearest McDonald's, after which they had turned onto deserted Elm Avenue. They saw nothing unusual until they reached the alley between the drugstore and the old garage.

It was a narrow passage, no more than eight feet across, and it extended into the darkness for about two hundred feet. Used tires were stored there, next to a garbage bin overflowing with discarded trash and old newspapers.

A rotting melon rind, a foul smell.

It was the sudden whirlwind that caught the boys' attention. It rose between the dirty brick walls. It picked up bits of paper, dust and pebbles, making a noise like a load of sand tumbling down a chute. Johnny stopped at the corner and watched in amazement.

Suddenly, there was a blinding crack of light in the center of the dark twister. The air became downright chilly, and a human shape appeared.

The boys shielded their eyes. When they looked again, a woman was walking towards them on unsteady legs, silhouetted against a purple glow that filled the alley behind her.

Spooked, Johnny Harder gave his bike a big push. He was followed by his companions, and they hightailed it out of there without ever looking back.

Rachel Rand held her head, shaking off a feeling of grogginess. She moved unsteadily down Main Street, surprised that her legs could carry her, and wondering where on Earth she was. At last she spotted a phone booth.

* * *

The phone rang in the darkened office that Peter shared with Thomas Archer. The only light came from the television set, which was still showing the ancient sci-fi movie on the BGS channel.

Chad Burke was passing by, whistling a tune that had been popular in Australia. He stopped to answer the call.

"*Inside Story.*"

"I'm looking for Peter Keller," the woman said hastily.

He caught the shaking in her voice.

"Yeah, well, he's not here, lady. He stepped out to grab a bite. Are you alright?"

"Tell him . . . tell him Rachel Rand called. Tell him I'm back."

"You're back?" mumbled Chad in astonishment. "You're not dead?"

"I wouldn't be calling you if I were dead."

"Stranger things have happened in this office, honey," Chad said cheerfully. "What else you want me to tell him?"

"Tell him I'm in Milton, Pennsylvania," she answered urgently. "And I'm leaving right now for New York. Tell him I'm coming over to see him."

The line went silent. Chad frowned, then shrugged and wrote down the message on a memo pad.

* * *

Rachel Rand's rental car raced past a sign on the open highway. She read:

NEW YORK STATE LINE—57 MILES.

She was concentrating on the dark road. She would be in Manhattan in a little over an hour. She wondered what Pete's reaction would be when he saw her again. She visualized his surprise. In the scene that went through her imagination, he would fold her into his arms. She would be able to entrust her knowledge to another human being at last. They would sit down in a cozy downtown bar, and she would tell him what she had discovered.

She felt strengthened by that thought. She lingered in the warmth of it.

A brilliant light abruptly became visible ahead of her. It was white, like the flash of a camera; then it changed color to a bewildering spectrum of intense reds and blues and greens. Her car slowed down almost immediately. The engine sputtered and died. She became panicky, dreading what was about to happen, still wildly hoping to escape.

"No!!! I won't let you touch me!" she yelled at the light, hitting the steering wheel with her fists.

"Go away!" she screamed.

But the huge saucer was already at ground level, swaying in the wind, hovering ahead of her, and backlit little figures were moving through the luminous fog, walking towards the car.

A surge of adrenaline flooded Rachel's whole body.

She fumbled, grabbing objects within her reach, frantically looking for something with which she could defend herself. But her muscles seemed to encounter an odd resistance. She felt like a grotesque living doll. She found it increasingly hard to move, as if she were fighting under water.

Her fingers became all tingly, as if she had dipped both hands into an anthill. Now the ants were crawling up her wrists, her elbows.

She was sliding, gliding, pulsing, sinking.

Rachel felt a familiar paralysis engulf her. First her hands, then her arms, then her neck, her legs ... The door was pulled opened on the driver's side. She staggered and bumped into one of the small gray beings. She lost consciousness.

They dragged her into an examination area within a gleaming white, evenly illuminated dome filled with strange instruments. They looked at her through their huge black eyes. A hard, cold finger moved along her arm, looking for a vein. Soon she was in a state suitable for the operation. The short gray aliens moved cautiously around her body. Two of them stood behind her, doing something to her head, as if prepping her for some kind of invasive surgery. They restrained her arms and her skull. A long, needle-like device entered the corner of her right eye, aimed very precisely at the frontal lobe of her brain.

* * *

The two aliens left Rachel in the examination area and entered an adjoining room, which looked much more terrestrial than the saucer's main dome. Colonel Trent, the man in black, Joe Wilkinson, Dr. Beverly Bernard and the Beard were there, sitting around, calmly discussing what they would henceforth call 'The Rachel Episode'."

"It was only an anterior capsulatomy," Dr. Bernard explained to her teammates. "She'll recover very nicely."

"Excuse me?" asked the Beard. "You're getting technical again."

"A simple form of lobotomy," she said curtly.

"Why did you feel that surgery was the method of choice?" Wilkinson asked. "We've perfected reinforcement schedules that serve the same purpose."

"We didn't have time for a full course of depatterning. It won't affect her a great deal; she'll just have to go through life with a little sense of confusion."

Colonel Trent nodded. "That's good," he said. "She's a nice girl, and she's served her purpose very well, as an abductee." He added with a zest of irony: "It would be a shame to lose her, especially now that the New Age is upon us. She deserves to enjoy it."

Joe Wilkinson jumped back into the conversation.

"In any case she's no threat to Alintel, from here on," he remarked. "But what about Borodine?"

"It's all under control. We've resumed the voices in his head," said Beverly. "And I've got him back on his regular medication."

The man in black nodded, satisfied. He walked over to the aliens and patted one of them familiarly on the back.

"Good work. Sit down, cool off," he said. "I'll call General Bushnell to report. He'll be glad to know that this phase of the operation is over."

"When is Peter Keller's tape running on *Inside Story?"* asked Beverly Bernard, methodically placing her surgical instruments back into a shiny metal case.

"Tomorrow. The whole country will be watching."

"That's good," said one of the little aliens, who sat down, crossed his legs and lit a cigarette. "Maybe I'll get some time off. I haven't had a fucking vacation in three years."

The other alien stretched grotesquely and held his misshapen head in both hands. Then, with a single smooth motion, the dwarf removed his alien mask.

* * *

It was the light that was the most depressing, the dull, dusty light falling on the cracked pavement, on the green-grey walls that hadn't been painted in twenty years. It was Friday night, so the corridors of Bellevue Hospital in Manhattan were crowded with families carrying flowers, orderlies pushing carts, a nurse rushing somewhere, and groups of visitors who, like Keller, were trying to read the signs and cryptic directions to the various psychiatric wards.

There were patients walking around, too, wrapping indecent green robes as tightly as they could around their frail bodies and dragging their paper shoes over the ceramic tiles that smelled of disinfectant. Peter realized for the first time in his life how vague and pliable the boundary was between the world of the healthy, to which he was presumably still attached, and the world of the sick and confused, the world where Rachel Rand now belonged.

He had nervously anticipated what it would be like to see her at the hospital. After getting her message, he had waited for her in vain for most of the night.

At five A.M. the Highway Patrol, in response to his repeated calls, had told him what little they knew: A car she had rented had been found abandoned by the side of a road in New Jersey. Later the police had spoken to the owner of a 7-11 store, who was puzzled by an incoherent female customer who matched her description. A patrol car was dispatched. She opposed no resistance. They sent her to an emergency psychiatric service for evaluation, and she ended up at Bellevue.

At first Peter thought she didn't even recognize him. She was sitting on the edge of her bed, dressed in the same blue sweater she had been wearing at the Four Corners, and she was staring past him as if he were invisible.

"Hi, Rachel . . ." He began softly, careful to keep his emotions in check, careful not to startle her.

The sound of his voice seemed to trigger an awareness in her. She tried to focus on his face. She brushed her hair from her forehead, got up slowly and took a step towards him, hesitating, her eyes suddenly filling up with tears.

He held her against him, sobbing, her arms around his neck. She didn't say a word.

"Where were you? What happened to you?" he finally asked softly.

She hesitated: "I went on a trip?" she said, more as an interrogation than as a statement. "Maybe it was a dream. You were there. A long trip, faraway. Then you weren't there any more."

"You disappeared," Peter stated. "Where did you go?"

She thought it over. "Too much light," she said. "Bright light . . . hurt me bad. A big disaster's about to happen. Very sad." More tears came to her eyes, streamed down her cheek. Peter wiped them off. "So I came back here."

He looked at her face. She was tense, trying hard to remember, searching for words to match memory that eluded her mind. She managed a little smile of apology.

"Do you remember why we went on this trip, you and I?" asked Peter.

A cloud seemed to obscure her eyes. She frowned and simply replied: "I don't want to go on another trip, ever."

She buried her face in his shoulder, and they were both very quiet for a long time.

THE END

ABOUT THE AUTHOR

Photo © 1995 by Yves Bosson

A computer scientist by profession, Dr. Jacques Vallee was born in France, where he was trained in astrophysics. He was awarded the Jules Verne Prize for his first novel in French, entitled *Sub-Space* and written at age nineteen. He moved to the United States, receiving his Ph.D. in computer science in 1967 from Northwestern University, where he was a close associate of the late Dr. J. Allen Hynek. He eventually published three more novels in French. *Fastwalker* is his first work of fiction in the English language.

The author of many scientific articles and of three books about high technology, Dr. Vallee first became interested in the subject of UFOs when he witnessed the destruction of tracking tapes of unknown objects at Paris Observatory. He was the first to compile a computer catalogue of UFO landings, and he was also the first to document the striking similarities between descriptions of encounters in contemporary abductions and medieval reports of elves, fairies and demons in folklore. His research into the phenomenon has taken him to many places in the United States and to many countries around the world, from France to Scotland and from Australia to Russia and Brazil.

His unique approach was crystallized when he served as the real-life model for the character of the French scientist played by Francois Truffaut in Steven Spielberg's classic film *Close Encounters of the Third Kind*. His books have been translated into all the major languages and represent millions of copies in print. His most recent work, *Forbidden Science*, was published in hardcover in 1992 by North Atlantic Books.